D0929976

Series Editors: Jack Sutcliffe
Peter Woodcock

Marketing Decisions: A Bayesian Approach

BEN M. ENIS
University of Houston

CHARLES L. BROOME
East Carolina University

Intertext Marketing Research Series

Published by
International Textbook Company Limited
24 Market Square, Aylesbury, Bucks., HP20 1TL

First published in the USA 1971 by
Intext Educational Publishers, 257 Park Avenue South, New York

ISBN 0 7002 0202 0

PRINTED IN GREAT BRITAIN BY OFFSET LITHOGRAPHY BY BILLING AND SONS LTD., GUILDFORD AND LONDON

Series Preface

This series will provide for the need for marketing research texts which draw both on American and on European material. Further, we believe these small books will be a valuable alternative in marketing research reading. If marketing research is to be covered in one volume, that volume will of necessity be large, and some people find that to consult a tome is off-putting. Our approach has been to break marketing research into its components and to obtain from different authors a small book about each part.

We aim to provide our readers not only with books which are easy to handle, but also with books which deal in some depth with particular aspects. Additionally, each book is sufficiently well referenced to enable both the student and the practitioner to explore any special feature more fully.

The series is directed firstly towards students and teachers of marketing and marketing research at Universities and Polytechnics. Secondly, it is intended to meet the particular needs of marketing research practitioners. It is not by chance that our authors either are, or have been practitioners, and we examine the theoretical background and the practical problem together.

Thirdly, we anticipate that certain volumes will be of interest to those working in other areas of social research. We believe that problems of, for example, attitude research are very similar whether we are considering attitudes to groceries or to our fellow citizens. Problems of questionnaire design, the interview situation, measurement of attitudes and behaviour, the analysis of findings, are essentially similar irrespective of the origin of the inquiry, or the use to which the results are put. Differences between marketing research and other forms of social research relate more to the content than to the methodology, and are thus superficial rather than profound. In practice, sociologists and psychologists are as likely to be interested in the findings of advertising motivation, pricing and distribution research as are marketing men. To underline our belief in the wide-ranging relevance of research methods, we have not hesitated to use examples from many backgrounds. Indeed, the necessity for keeping a wide view has become increasingly apparent as work on the series has progressed.

Lastly, and by no means least, we hope that the non-specialist will gain an appreciation of what this sort of research is setting out to do. If we have managed to convey any feeling to the manager outside those facets of industry mainly associated with this sort of research, that the approaches we discuss might be applicable to his problem areas—that there might be something in it for him—this will be reward in itself for the editors, authors and publishers of the series.

1973

J. SUTCLIFFE
P. WOODCOCK

Preface

In writing a foreword to the Intertext Marketing Research Series edition of this book, certain aspects of the Bayesian approach might be discussed with profit.

First of all, what is it that distinguishes the *Bayesian* approach from other possible approaches? The immediate answer to this is that Bayesians specifically and explicitly demand that the decision maker use his judgement and his experience *and* the judgement and experience of those he chooses to consult. This is in direct contrast to the approach of *classical statistics* which appears to be independent of *subjective* judgement.

This leads on to a second, and consequent difference, namely that Bayesians ask different sorts of questions. If, for example, we take Hypothesis Testing, the classical statistician is concerned with the risks of making Type I and Type II errors. The Bayesian, on the other hand, is concerned with the costs associated with these errors. In general terms, Bayesians ask questions about the cost and value of what they are doing, whether it be designing a sample, testing an hypothesis or deciding whether to gather additional information.

As a second question, we should ask how successful the Bayesian approach has been in practice. In the recent past, two review articles have appeared describing U.S. (Brown) and U.K. (Longbottom) experience. The list of companies who have tried this approach is impressive. The problems they have unearthed are considerable, but not overwhelming. None of the problems is associated with an inadequate theoretical basis: all are concerned with the practical applications. How, for instance, can practical guidelines be developed for the assessment of probabilities? Where should those responsible for processing find themselves in the overall structure of the firm? How does one overcome the human problem of managers feeling themselves threatened?

Thirdly, what advantages are claimed for this approach? The most important is that assumptions are forced out into the open—one can see immediately what has and has not been taken into consideration. This presents the decision maker with a useful vehicle for communication—the reasoning behind any decision becomes apparent to those without an intimate knowledge of the details. By increasing the openness of the decision process, we are presented with an opportunity to improve the quality of our decisions. By highlighting those factors to which the decision is particularly sensitive, we point up areas where it may well be worth our while to gather additional information prior to coming to a decision, as well as providing ourselves with estimates of the cost of delaying our decision and the potential value of any additional information we might gather.

Lastly, we should look at a development of particular interest to marketing researchers, the evaluation of research findings. Two articles and a book, by Professor Rex V. Brown, are referred to in the text. By taking a Bayesian approach, Brown suggests that we can put 'managerial' values on research findings. By subjectively assessing various sources of error, he suggests that we can come to conclusions about the credibility of our findings which are at once both more useful and more comprehensible to those who have to act on them. In turn, we can develop ways of assessing alternative research designs, and Brown's suggestions help to resolve those regular difficult problems of deciding whether for example, to use personal interviews, telephone interviews or mail questionnaires. For the practicing researcher here, at last, is a methodology of choice. It is of some interest to learn that Brown did his development work whilst employed by Martech Consultants Ltd., in London as a marketing researcher. One presumes, though he does not say so, that his work grew out of the day-to-day problems with which he found himself faced whilst doing practical commercial research.

What, then, is the state of the art? One quickly becomes aware that the large companies—ICI, Ford, DuPont, General Electric, Unilever—have tried this approach and that some of them are pleased with the results. What is much less clear is the importance of its impact on business as a whole. So far, only a tiny fraction of companies have even got as far as experimenting. Will this approach become as important to management as mathematics has become to engineering? Will it merely remain on the fringe of decision making, to be used occasionally, but in no sense to be part of the mainstream? The answers to these questions are clearly of major importance to businessmen, and are largely in their hands. The ultimate outcome is likely to depend on how interested they are to improve the quality of their decisions.

Bibliographical Appendix

For those not familiar with the basic concepts, we would recommend the following:

SCHLAIFER, ROBERT, *Probability and Statistics for Business Decisions*, McGraw–Hill, 1959.

This book has already achieved the status of a classic in its field. Schlaifer is regarded as one of the leaders of the Bayesian movement, and this book translates Bayes' two hundred year-old theoretical paper into practical terms. Schlaifer developed a set of monographs and tables which have cut out a great deal of tedious arithmetic, and enable approximate solutions to be arrived at quickly. He followed this up with another book:

Analysis of Decisions under Uncertainty, McGraw–Hill, 1969.

This examines a wide range of practical business problems, and can serve as a model for those wishing to try their hand at applying the theory to their own situations. Thirdly we would strongly recommend:

RAIFFA, HOWARD, *Decision Analysis*, Addison–Wesley, 1968.

This book is based on a series of lectures given by Professor Raiffa at the Harvard Business School, were both he and Schlaifer hold chairs. No mathematical skill beyond basic arithmetic is required, and Raiffa explores some of the underlying assumptions of this and other classes of statistics.

THOMAS, HOWARD, *Decision Theory and the Manager*, The Times Management Library, Pitman, 1972.
BYRNES, W. G. and CHESTERTON, B. K., *Decisions, Strategies and New Ventures*, George Allen & Unwin, 1973.

Actual case studies on use of decision analysis, drawing on the authors' experiences in the Economics and Statistics Department at Unilever.

ADDITIONAL JOURNAL ARTICLES

BROWN, REX V., 'Do managers find decision theory useful?' *Harvard Business Review*, pp78–89, May–June, 1970.

COPELAND, B. R., 'Statistical decision theory,' *Management Services*, Vol 5, pp45–51, May, 1968.

ENIS, BEN M., 'Bayesian approach to ad budgets,' *Journal of Advertising Research*, Vol 12, pp13–19, February, 1972.

FREDERICK, D. G., 'Industrial pricing decisions using Bayesian multivariate analysis,' *Journal of Marketing Research*, Vol 8, pp199–203, December, 1971.

HARRISON, P. J. and STEVENS, C. F., 'Bayesian approach to short-term forecasting,' *Operational Research Quarterly*, Vol 22, pp341–362, December, 1971.

INDUSTRIAL MARKETING RESEARCH ASSOCIATION, 'Probabilistic decision-making in practice—proceedings of a seminar,' *Journal of the Industrial Marketing Research Association*, Vol 7, No 3, August, 1971.

LONGBOTTOM, D. A., 'Decision making—a review,' *Management Decision*, Vol 10, pp224–42, Winter, 1972.

PARSONS, J. A., 'Statistical decision making,' *Journal of Systems Management*, Vol 23, pp44–5, June, 1972.

PHILLIPS J. D. and DAWSON, L. E., 'Bayesian statistics in retail inventory management,' *Journal of Retailing*, Vol 44, pp27–34, Summer, 1968.

Contents

Marketing Decisions: A Bayesian Approach

CHAPTER 1

The Nature of Marketing Decisions

According to one commonly accepted definition, *marketing* is

> the analyzing, organizing, planning, and controlling of the firm's customer-impinging resources, policies, and activities with a view to satisfying the needs and wants of chosen customer groups at a profit.[1]

This definition describes a complex, dynamic, science/art which requires that marketing managers make innumerable decisions, many of which are of significant consequence to their firms. For this reason, increasing attention is being paid to concepts and techniques that strengthen the efficiency of the decision-making process for marketing management. The authors believe that Bayesian analysis is such a tool. Consequently, the purposes of this book are to present the fundamentals of Bayesian analysis and to acquaint the reader with Bayesian applications in marketing management.

The present chapter provides an orientation to the Bayesian approach to marketing decision making. First we review the major internal and external dimensions of marketing decisions from the managerial point of view. From this discussion is derived a set of requirements which an analytical procedure for decision making in marketing should encompass. These require-

[1] Philip Kotler, *Marketing Management: Analysis, Planning and Control* (Englewood Cliffs, N.J.: Prentice-Hall, Inc., 1967), p. 12.

ments are then formalized in a generalized decision process which demonstrates, in outline form, how Bayesian analysis meets these requirements. The final section summarizes the chapter and thus serves as an overview of the remainder of the book.

DIMENSIONS OF MARKETING DECISIONS

In our discussion of marketing decision making, it is useful to differentiate between "decision variables" and "environmental variables." Decision variables are those factors within the firm which are relevant to the situation in question. Examples are product line policy and pricing strategy. The firm has some degree of control over these factors. Environmental variables, on the other hand, are factors which affect the outcome of marketing decision situations, but are beyond the direct control of the firm. Environmental variables include buyer behavior, economic conditions, legislative and judicial decisions, social institutions, and competitors' actions. Of course, the firm can influence environmental variables. Changing consumer behavior, for example, is a major objective of marketing.

Since the function of marketing is to satisfy the needs of buyers (both industrial and household consumers), these needs must be determined. That is, *demand* for the firm's products must be estimated. The problem of forecasting product demand is a complex one. A knowledge of economic considerations (income, general economic activity, efficiency of product performance) is necessary but not sufficient. Attention must also be given to such factors as cultural norms, individual preferences, and habit patterns. Such factors as competitors' actions, governmental regulations, geography, climate, and social institutions also influence the demand for a given product. The relationship between any one of these variables and demand is likely to be complex; moreover, the variables interact, and any of these influences is subject to change over time. In addition, information concerning environmental variables is often difficult to obtain and interpret.

Consequently, the marketing manager must make decisions in an environment which is complex, dynamic, and largely beyond his control. Yet he must be cognizant of the effect of such environmental variables upon his decision. Clearly, a systematic approach to problem solving in such an environment could be extremely valuable.

Within the firm, the marketing manager must consider a number of additional factors. He has some degree of control over these intrafirm factors, but still must analyze and coordinate their impact upon a particular decision. Marketing decisions are typically subdivided into four areas. *Product decisions* include such considerations as additions to and deletions from the product line, design of a particular product for maximum consumer utility, styling, and packaging. *Pricing decisions* concern setting price policies, price lines and specific prices. *Promotion decisions* include advertising copy, media selection, training and supervision of salesmen, and use of point-of-purchase material. *Channel decisions* concern such problems as selection of channels, supplier-vendor relations, and physical distribution.

The above subdivision of the responsibilities of marketing management is to some extent arbitrary. Decisions in these areas often interact. Product quality affects price, channel decisions influence media selection, and so on. Moreover, decisions by marketing managers affect and are affected by decisions made by managers of production, finance, and personnel. Product design and manufacture is in the province of the production manager. Budgetary constraints place limitations on marketing decisions. And personnel selection, training, and supervision problems are of concern to the personnel manager as well as to his marketing counterpart.

The above discussion illustrates some of the difficulties involved in applying formal analytical procedures to marketing decision situations. There are essentially three causes of these difficulties. First, many marketing problems are more or less unique. A new product, for example, is offered to the market only once. There are no historical data from which long-run relative frequencies can be calculated for traditional statistical analysis. Consequently, past experience often provides only general guidelines for decision making. Secondly, marketing's interface with buyers results in problems not found in other areas of the business. The buyer can think for himself. He may, therefore, change his mind, reorder his priorities, or otherwise alter his behavior patterns. By contrast, the production manager may feel reasonably confident that his raw materials will not refuse to become part of the product he is manufacturing.

A third barrier to formal analysis of marketing problems is the complexity of such problems. Consider, for example, the cost accountant's dilemma. The accountant would like to calculate the profit (or loss) associated with each marketing activity, as he does for activities in the production process. If more expensive raw materials are used in a given product, the accountant knows that the cost of manufacturing one unit of that product will increase. But if additional advertising dollars are allocated to that product, per unit costs may either increase (if sales increase less than proportionally) or decrease (if the sales increase more than proportionally). Of course, this comment does not do justice to problems of production cost accounting, but the point is that, for marketing costs, such problems are likely to be more numerous and more severe.

As Kotler remarks, "marketing decisions must be made in the context of insufficient information about processes that are dynamic, nonlinear, lagged, stochastic, interactive, and downright difficult."[2] More specifically, marketing decisions must be made about complex situations, which have significant impact upon organizational objectives, under conditions of uncertainty about potential outcomes, with considerably less than perfect knowledge of the environment, and little control over many of the relevant variables.

Marketing managers have, of course, been handling such situations with varying degrees of success for a long time. Many develop, through observation, experience, and trial-and-error, a sound intuitive "feel" for certain marketing situations. They integrate information (subjective, historical, or experimental) with their intuition to arrive at an "adequate" decision.

This approach, while certainly not without merit, suffers from two basic weaknesses. First, the manager who uses this intuitive procedure cannot be sure that he has considered all of the factors relevant to the situation in question, nor that he has evaluated the logical consistency and possible interaction of these factors. Consequently, important data or relationships may be overlooked or misinterpreted. The second weakness of

[2] Kotler, *ibid.*, p. vii.

this approach is that it is essentially personal. The decision maker cannot explain, either to superiors or to subordinates, how he reached his conclusions. Only the results of the decision, not the decision *process*, are available for review. The superior cannot evaluate the manager's decision procedure, nor can the subordinate learn from it.

A formalized approach to decision making can overcome these weaknesses. An explicitly structured decision process will include all relevant factors, does require logical consistency of reasoning, and can be communicated. The next section outlines the decision process to be followed in this book.

THE BAYESIAN DECISION PROCESS

Let us define a *decision* as a choice among alternatives for the purpose of achieving some specific objective. This definition is somewhat imprecise, but it does indicate that an operational procedure for efficiently making choices among alternatives can be constructed. This procedure has many names (e.g., the scientific method, the problem-solving approach) and has been formulated by various decision theorists in basically similar steps. We shall consider the decision process in Bayesian terminology. The steps are problem definition, prior analysis, preposterior analysis, additional data assembly, posterior analysis, and selecting a course of action. This procedure is illustrated in Fig. 1-1.

The first step in reaching a decision is carefully to identify the dimensions of the problem. The decision maker or manager must know the objectives he is trying to accomplish, the constraints which delineate acceptable solutions, alternative courses of action available, and the criteria used in choosing among the alternatives. To be meaningful, each of these factors must be operationally delineated, preferably in quantitative terms. In marketing, objectives can often be stated in terms of market shares or profits, constraints in terms of costs and time, alternative courses of action in market-share percentages, discounted cash values, and so on.

Once the problem has been delineated, in terms of objectives, constraints, and alternative courses of action, the decision maker can logically analyze the problem. On the basis of his prior knowledge and experience plus the facts at hand, he may

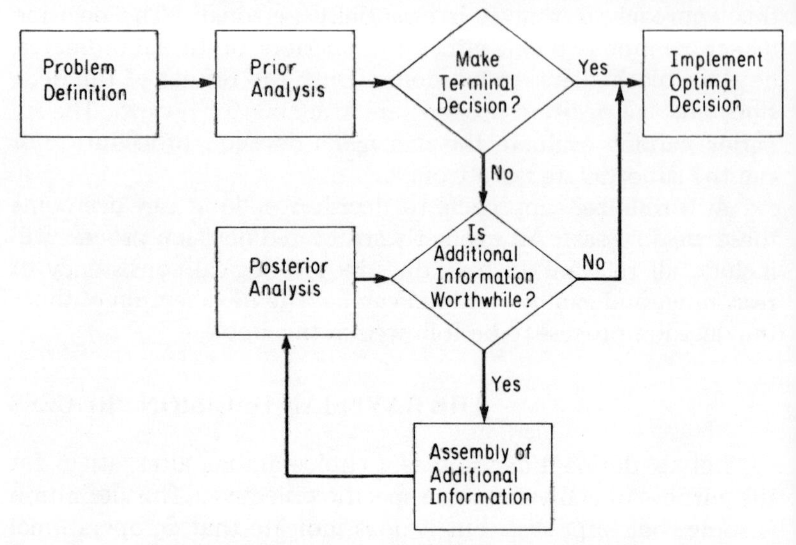

Fig. 1-1. A Bayesian view of the decision process.

be sufficiently confident in his ability to predict the outcome of each alternative that he will select the one that best accomplishes his objectives within the constraints. In Bayesian terms, this selection is called a *terminal decision*. If he is less certain of the various outcomes, the decision maker might prefer to obtain additional information before making the terminal decision.

Information, however, is seldom a free good. There is usually a cost, in time and resources, associated with the collection of data. Consequently, the decision maker must first determine whether additional information would be worthwhile. That is, he would like to know whether the reduction in uncertainty about the outcomes would more than offset the cost of collecting the data. This step in the decision process is termed *preposterior* analysis. If additional information costs more than it is worth, the terminal decision is based on the prior analysis.

If on the other hand the additional information is worth more than it costs, it is assembled. Data-collection procedures which might be employed include search of secondary sources, observation of the situation, surveys of persons involved, sampling of elements of the problem, experimentation with key variables, and simulation of the situation. The data collected must then be integrated with the prior information.

The integration and evaluation of "prior" and "additional" information is termed *posterior* analysis. As in the prior analysis, the purpose of posterior analysis is to select the course of action which best accomplishes the objectives of the problem within the stated constraints. This purpose implies that the decision maker or manager has a decision framework: a set of values and a procedure for evaluating alternatives in terms of these values. We shall explore this topic in some detail in Chapter 3 and propose a basic model for decision making.

Once the decision maker chooses a course of action, he must implement it or influence the organization to do so. Otherwise, his "decision making" will have no practical significance. A complete discussion of the process of decision implementation is beyond the scope of this book, but the reader should keep in mind the necessity of this step.

Bayesian analysis provides the means for making this decision process operational for marketing problems. It provides a means for evaluating complex problems under conditions of uncertainty, and does so in a manner which is comprehensible to the decision maker with relatively little training in probability and statistics. Moreover, the use of Bayesian procedures allows communication of the decision process.

In summary, Bayesian analysis is essentially characterized by five elements:

1. The decision maker is involved in a situation in which there are at least two alternative ways of reaching a specified objective(s), and he has the power to decide among the alternatives.
2. The decision maker is uncertain as to which decision alternative to select, because he does not know the set of environmental conditions (state of nature) which will actually prevail at the time the decision is implemented.
3. The decision maker has some knowledge of the situation, e.g., relevant payoffs of alternatives, and likelihood of occurrence of various events or states of nature which affect these payoffs.

4. The decision maker is willing to use expected value as his decision criterion.
5. The decision maker may be able to obtain additional information (at some cost) which might change his assessment of the situation.

Consequently the decision maker can incorporate whatever information (subjective, historical, or experimental) he has within a systematic framework for decision making in complex, uncertain situations.

OVERVIEW OF THE BOOK

We have endeavored to present the fundamentals of Bayesian analysis as it applies to marketing decisions. This chapter has reviewed the decision process, described the environment within which marketing decisions are made, and outlined the methodology of Bayesian analysis. Since Bayesian analysis employs probability theory, and does so in a manner which some experts consider controversial, we have devoted Chapter 2 to a review of the rudiments of probability theory.

Chapter 3 is concerned with choice criteria—determining the basis upon which alternative courses of action should be evaluated. Three types of decision environments are considered: certainty, risk, and uncertainty. Under uncertainty (i.e., when the decision maker does not know the probability of occurrence of each of the possible states of nature), expected value criteria constitute the most meaningful bases for selection of the best course of action. Chapter 4 presents the elementary concepts of Bayesian analysis. The payoff matrix is constructed, prior analysis is performed, and the value of additional information is calculated.

Chapter 5 extends the concepts presented in Chapter 4 to include the binomial distribution, and posterior analysis. Chapter 6 discusses the role of the normal distribution in prior, preposterior, and posterior Bayesian analysis. Chapter 7 is an evaluation of Bayesian analysis from a managerial perspective. The Bayesian approach is summarized and contrasted with traditional statistical approaches to marketing decisions, the limitations of the Bayesian approach are reviewed, several real-world applications are discussed, and future developments in the use of Bayesian analysis are predicted.

REFERENCES

Alderson, Wroe and Paul E. Green. *Planning and Problem Solving in Marketing.* Homewood, Ill.; Richard D. Irwin, Inc., 1964, Part I.

Part I of this book is devoted to the marketing function in organizations, and to a general discussion of problem solving techniques.

Kotler, Philip. *Marketing Management, Analysis Planning and Control.* Englewood Cliffs, N.J., Prentice-Hall, Inc., 1967.

Best general treatment available, in the authors' opinion, of the field of marketing management.

Roberts, Harry V. "Bayesian Statistics in Marketing," *Journal of Marketing*, January 1963.

A nontechnical introduction to Bayesian analysis in a marketing context.

CHAPTER 2

Rudiments of Probability Theory

The decision maker faces uncertainty. If he knew which state of nature would actually occur at the time his decision was to be implemented, he could select the course of action that would optimize his objective for that state of nature. Since he often does not know which state will occur, the manager must develop, implicitly or explicitly, a decision procedure for making choices under conditions of uncertainty. The Bayesian procedure consists of three steps. First, all states of nature to be considered are enumerated. Secondly, a probability of occurrence is assigned to each state. Thirdly, the expected value of each alternative is calculated.

This chapter provides foundation material for understanding and applying this procedure to marketing problems. The meaning of probability is discussed and computational rules are reviewed. This chapter also acquaints the reader with terminology and symbols to be used throughout the book.

THE MEANING OF PROBABILITY

Although the study of probability began earlier, it was in the seventeenth century that Pascal and Fermat formulated fundamental principles for the mathematical theory of probability.[1]

[1] A good summary of the historical development of probability theory can be found in Peter C. Fishburn, *Decisions and Value Theory* (New York: John Wiley & Sons, Inc., 1964), Chap. 5.

Laplace made substantial contributions to the theory during the eighteenth century. His definition of probability in terms of equally likely events was almost unchallenged until the 1920's when von Mises interpreted probabilities as long-run-relative frequencies. Since then, Ramsey, de Finetti, and Savage have been major exponents of the concept of personal or subjective probabilities. L. J. Savage's book, *Foundations of Statistics*, is considered the most important work on personal probabilities and provides the foundations for the use of subjective probabilities in modern decision theory.[2] Savage's work has been extended, with emphasis on business applications, by Robert Schlaifer.[3]

In addition to the terms used above to denote various concepts of probability, the reader may have already encountered some of the following terms: *objective*, *impersonal*, *inductive*, *empirical*, *intuitive*, and *logical probabilities*, *relative frequency*, *degree of conviction*, *degree of rational belief*, and *random chance*. These terms as well as others have been used by various authors to denote different meanings of probability. Still other authors have used different adjectives to denote the same meaning of probability. Such confusion about the meaning of probability led one well-known author to comment, "indeed, probability has been called the most important concept in current science, especially . . . as nobody has the slightest idea what it means."[4]

Although the exact meaning of probability is important to a philosophical treatment of the subject, it is adequate for our purposes to recognize that the many different interpretations tend to place probabilities in one of two classes: objective or subjective probabilities. In the objective class, probability may be interpreted as relative frequencies. In the subjective class, probability may be interpreted as strength of personal belief. These concepts are discussed in the following paragraphs.

[2] L. J. Savage, *Foundation of Statistics* (New York: John Wiley & Sons, Inc., 1954).

[3] Robert Schlaifer, *Probability and Statistics for Business Decisions* (New York: McGraw-Hill Book Company, 1959).

[4] E. T. Bell, *Mathematics: Queen and Servant of Science* (New York: McGraw-Hill Book Company, 1951), p. 377.

Objective Probabilities and Relative Frequencies

Consider the common experience of tossing a six-sided die. Before the toss, we are unable to predict which one of the six faces will be showing after the die is tossed. There is uncertainty as to which face will actually appear, so we term the event that a particular face will be showing a chance or random event.

Some insight can be gained by considering the question of why are we unable to predict the outcome of a tossed die. Presumably, the toss of the die follows laws of motion which are deterministic. That is, given that the die meets certain specifications (its sides form a perfect cube, its density is uniform, etc.), then the behavior on each toss should be the same. The outcome varies on each toss because of many small variations—for example, the position of the die before the toss, the velocity at the time of release, the density of the air surrounding the die, and the character of the surface upon which the die is thrown. Because we cannot predict accurately the conditions existing at the time of the toss, we cannot predict the outcome of any given toss. Since we cannot predict the outcome of a given toss in advance, we say that chance determines outcome.

Although the outcome of a single chance event cannot be predicted, experience indicates that regularities do occur in repetitions of chance events. For example, empirical studies have shown a wide variation in the percentage of times a given face will appear in a small number of tosses; as the number of tosses increases, the percentage of times a given face will appear becomes stable. Therefore, the percentage of times that we observe a particular outcome tends to become constant as the number of observations increases.

If we assume that the die-tossing experiment described above can be repeated indefinitely under identical conditions, we may apply the relative frequency concept of probability to the result of a given toss. Consider a particular outcome, such as two spots appearing on a single roll. Both experience and empirical evidence indicate that the percentage of times that a 2 occurs will approach a unique number as the number of trials is increased indefinitely. Thus the probability that a 2 occurs can be thought of as the long-run relative frequency that a 2 occurs. Hence, the number which we call probability gives a numerical description of a chance event.

Although we cannot predict the specific outcome of a chance event, it helps to be able to predict what will happen if the experiment is repeated a large number of times. There are many real-world situations analogous to repetitive experiments with chance outcomes. In these situations, the decision maker frequently wants to make decisions in such a way as to obtain optimal long-run results. Knowledge of the probabilities of each possible outcome of a situation is very useful. For example, knowledge about deaths of various segments of the population enables insurance companies to establish life insurance premiums for individuals based on present age, occupation, and life expectancy. Similarly, knowledge of probabilities is embodied in designing quality control systems, flood-control facilities, employee pension plans, and developing many other decision rules.

Subjective Probabilities and Degree of Rational Belief

Decision makers, especially in marketing, are sometimes faced with unique problems, i.e., problems which have not previously occurred and are unlikely to occur again. If the decision maker is unable to predict the outcome of such unique situations, he may find it convenient to view the outcome as a chance event. The probability concept permits quantitative estimates of the uncertainty associated with such problems.

In fact, as commonly understood, the word *probability* often refers to unique situations. Examples are statements such as: "the probability that a democrat will be elected president in the next election is .47," and "the probability that State University will win the football game Saturday is .62." Since the environmental factors which determine the outcome cannot be repeated, the associated probabilities cannot be interpreted as relative frequencies. Instead, we say that probability is being used to express one's degree of rational belief. Probabilities assigned in this manner are often referred to as *personal or subjective probabilities.* Decision makers often have strong intuitive feelings about the likelihood of a particular outcome that is determined by chance. The assignment of subjective probabilities to such outcomes provides the decision maker an opportunity to quantify his feelings about them.

In statistical decision theory, both subjective and objective probabilities play an important role. One of the fundamental assumptions underlying the remainder of this book is that decision makers can quantify their feelings about particular situations by assigning probabilities to uncertain events. The assignment of probabilities is subjective in the sense that two individuals with different backgrounds and knowledge about a situation will probably not assign the same numerical value to a given outcome. On the other hand, it seems reasonable to expect that two individuals with similar backgrounds and knowledge about a situation would assign probabilities having approximately the same numerical value to a given outcome.

In summary, the philosophical interpretation of objective probabilities and subjective probabilities is quite different. The former concept is based on a large number of repetitions of an experiment while the latter pertains to an experiment that can be performed only once. Insofar as practical problems are concerned, these two situations appear to be limiting cases, i.e., knowledge about most problems seems to lie somewhere between the extremes of hard long-run data and complete ignorance. Some historical information will usually be available, but almost never as much as the decision maker would like to have. Thus some personal judgment will of necessity be incorporated into the assignment of probabilities. For this reason it is fortunate that the mathematical basis for making computations involving probabilities does not distinguish between subjective and objective probabilities. The following sections briefly review pertinent aspects of probability theory.

THE EVENT SET

Let us continue our interest in determining which face of a fair die appears on a given toss. We first need to identify all the outcomes, or more formally, enumerate the logical possibilities, in order that we may make probability statements about them. In tossing a die, we are considering an experiment which has six logical possibilities. These six possibilities may be referred to as the event set E. Each logical possibility is a simple event and may be indicated by $(e_1, e_2, \ldots, e_6)$ where e_1 is the event that one spot appears, e_2 denotes the appearance of two spots, and so on. These simple events are *mutually exclusive* and *collectively exhaustive*. By mutually exclusive is meant that one and only one simple event can occur when the experiment is performed. By collectively exhaustive is meant that only the simple events listed can occur. The relationship between the event set and the simple events of the above experiment may be shown as follows: $E = (e_1, e_2, \ldots, e_6)$. This statement is read, "the event set E has six outcomes, denoted e_1, through e_6." Since the outcome of a given toss is uncertain, the event e_i, $(i = 1, \ldots, 6)$ is termed a random variable.

We should note that there are other ways of viewing the die-tossing experiment. For example, we might be interested only in the occurrence of an odd or even face. If so, we would have only two simple events; this relationship may be shown as follows:

$$E = (E_1, E_2)$$

where

$$E_1 = (e_1, e_3, e_5)$$

$$E_2 = (e_2, e_4, e_6)$$

There are many other ways of analyzing this experiment; each would require that we define our simple events in a different way. Usually, the context of a particular problem suggests an appropriate definition of simple events.

THE ELEMENTARY PROBABILITY MODEL

The axiomatic approach to the development of probability theory assumes that the probabilities assigned to events are known; it is concerned with computations related to them.[5] For our die-tossing experiment, let us denote the probability of the occurrence of simple event e_i $(i = 1, \ldots, 6)$ by $P(e_i)$. The relationship between $P(e_i)$ and e_i is called a *real-valued function.* The set e_i $(i = 1, \ldots, 6)$ is the domain of the function and the set $P(e_i)$ is its *range*. Thus the mathematical model of the experiment is determined by the event set e_i and the function $P(e_i)$. The system for manipulating this model is based upon

[5] The reader will recall that the probabilities may be "known" either objectively (on the basis of logic or historical evidence) or subjectively (degree of belief).

three axioms for assigning probabilities to simple events:

$$0 \leqslant P(e_i) \leqslant 1 \qquad \text{for all } e_i \tag{2-1}$$

$$\sum_{i=1}^{n} P(e_i) = 1 \tag{2-2}$$

where $P(e_i)$ is the probability of a simple event and there are n events, and

$$P(E_1 + E_2) = P(E_1) + P(E_2) \tag{2-3}$$

where E_1 and E_2 are mutually exclusive and represent the union of two or more simple events.

The first axiom states that a probability assigned to an event is a number between 0 and 1, including the end points. The second axiom states that the sum of the probabilities assigned to a set of mutually exclusive and collectively exhaustive events must equal 1. The third axiom enables us to compute probabilities involving combinations of simple events. For example, for $E_1 = (e_1, e_3, e_5)$ and $E_2 = (e_2, e_4, e_6)$, $P(E_1) + P(E_2) = 1$. Axiom 2 can then be applied to these composite events and thereby extend the use of probability theory to all subsets of events in a given experiment.

CALCULATION OF PROBABILITIES

When the simple events of an experiment can be enumerated, we can assign a number or "probability" to each event as follows:

$$P(E_1) = \frac{\text{Number of simple events in } E_1}{\text{Number of simple events in } E} = \frac{n(E_1)}{n(E)} \tag{2-4}$$

The symbol $P(E_1)$ is read "the probability of E_1" or "the probability that E_1 occurs." For example, in our die-tossing experiment, let E_1 be the event that a face with an odd number of spots appears on a given toss. The simple events in E_1 are e_1, e_3, and e_5; the simple events in E are e_1, e_2, e_3, e_4, e_5, and e_6. Then

$$P(E_1) = \frac{n(E_1)}{n(E)} = \frac{3}{6} = \frac{1}{2}$$

This result is the unconditional or marginal probability that E_1 occurs.

It is sometimes helpful to view the outcomes of an experiment as a sample space. Sample spaces can be represented by Venn diagrams. For example, let the six outcomes of the die-tossing experiment be represented by the six points shown in Fig. 2-1. The probability of E_1 then is the ratio of points in E_1 to the total number in the experiment E.

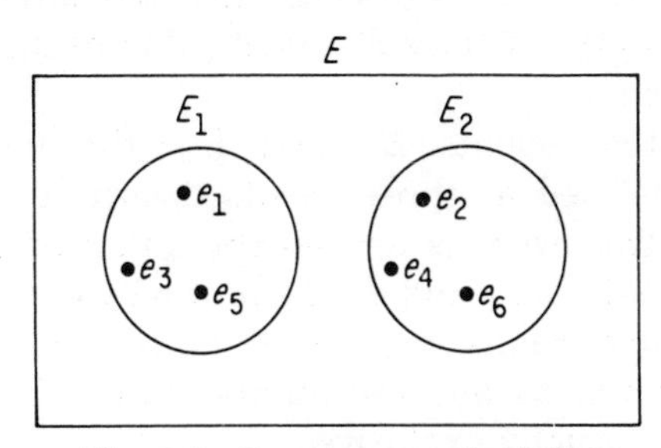

Fig. 2-1. Sample space depicting mutually exclusive events in die-tossing experiment.

Another observation can be made from Fig. 2-1. The probability of E_1 and the probability of *not* E_1, i.e., the probability that E_1 does not occur, denoted by $P(E_1')$, is a set of mutually exclusive and collectively exhaustive events. Since, by Axiom 2, the sets $P(E_1)$ and $P(E_1')$ total 1, the following formula results:

$$P(E_1') = 1 - P(E_1) \qquad (2\text{-}5)$$

In many instances, one of these probabilities $[P(E_1)$ or $P(E_1')]$ may be easier to compute than the other. Consequently, calculations may be simplified by recognizing that knowledge of $P(E_1')$ is equivalent to knowledge of $P(E_1)$.

The calculations involving probabilities may be further simplified by applying certain rules for combining them. The rules may be derived from the three axioms using the algebra of propositions.[6]

[6] Proofs are given in J. Freund, *Mathematical Statistics* (Englewood Cliffs, N.J.: Prentice-Hall, Inc., 1962) and A. M. Mood and F. A. Graybill, *Introduction to the Theory of Statistics.* 2d ed. (New York: McGraw-Hill Book Company, 1963).

Addition Rules

Mutually Exclusive Events. In our die-tossing experiment a given outcome excludes the possibility of the occurrence of any other simple event. Such events are said to be *mutually exclusive.* This idea can be formalized as follows. In Fig. 2-1 composite events E_1 and E_2 are mutually exclusive; if E_1 occurs, E_2 cannot; if E_2 occurs, E_1 cannot. If two events E_1 and E_2 are mutually exclusive, $(E_1 \cap E_2) = \phi$, i.e., the intersection of E_1 and E_2 is an empty set since E_1 and E_2 do not include any common simple events.

For any two events E_1 and E_2, the combined event $(E_1 \cup E_2)$ is defined as those events occurring in either E_1 or E_2, or both. This event is termed the union of E_1 and E_2. As shown in Fig. 2-1, if E_1 and E_2 are mutually exclusive, then the number of simple events in $(E_1 \cup E_2)$ is the sum of the number of simple events in E_1 and the number in E_2. Therefore, when E_1 and E_2 are mutually exclusive,

$$P(E_1 \cup E_2) = P(E_1) + P(E_2) \tag{2-6}$$

This formula is known as the *special rule* for addition of probabilities.

Compound Events. If E_1 and E_2 are not mutually exclusive events, then $P(E_1 \cup E_2)$ is something less than $P(E_1) + P(E_2)$. For example, consider again the toss of a die and define the events E_1 and E_2 as follows:

E_1: number of spots is odd

E_2: number of spots is greater than 3

The situation is depicted in Fig. 2-2.

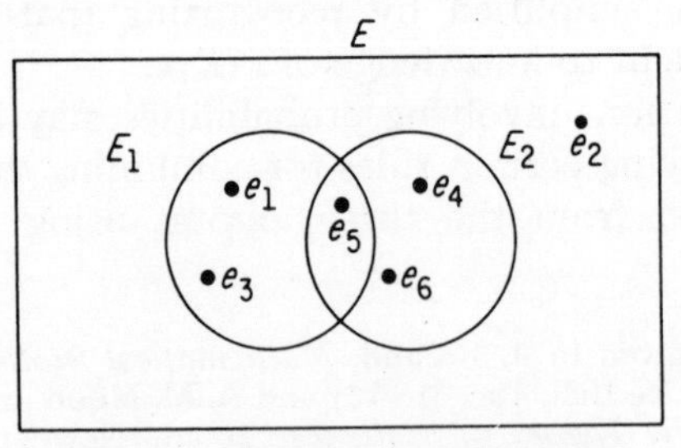

Fig. 2-2. Sample space depicting compound events in die-tossing experiment.

The simple event e_5 will be counted twice if we sum $P(E_1) + P(E_2)$. To eliminate double counting, the number of twice-counted simple events, $(E_1 \cap E_2)$, must be subtracted once. Thus for compound events,

$$P(E_1 \cup E_2) = P(E_1) + P(E_2) - P(E_1 \cap E_2) \qquad (2\text{-}7)$$

This equation is the *general rule* for addition of probabilities. Note that rule (2-6) is a special case of rule (2-7), since $P(E_1 \cap E_2) = 0$ for mutually exclusive events. Both rules can be readily extended for cases involving more than two events.

Multiplication Rules

Conditional Probabilities. Assume that a die is tossed and we are told that the number of spots is odd, but not the number of spots showing. We wish to know the probability that the number of spots showing is greater than 3. This situation is shown in Fig. 2-2, where again

E_1: number of spots is odd

E_2: number of spots is greater than 3

It is instructive to view the example as an experiment in which we have acquired "additional information" which enables us to reevaluate the logical possibilities for simple events. We are told that the number of spots is odd; this fact limits the possible simple events which can occur to those included in E_1. The probability that we are interested in is $P(E_2|E_1)$: the probability that the outcome is greater than 3 given that it is odd. This probability is termed a *conditional probability* and is defined as follows:

$$P(E_2|E_1) = \frac{P(E_2 \cap E_1)}{P(E_1)} \qquad (2\text{-}8)$$

where $P(E_2 \cap E_1)$ is the probability of the joint occurrence, i.e., the intersection of E_2 and E_1. The reader will note that Eq. 2-8, like Eq. 2-4 is a ratio of the probability of the event in question to the total probability. If any two of these probabilities are known, the third can be calculated using Eq. 2-8.

If we multiply both sides of Eq. 2-8 by $P(E_1)$, we have

$$P(E_2|E_1) \cdot P(E_1) = P(E_2 \cap E_1) \qquad (2\text{-}9)$$

And since $P(E_2 \cap E_1) = P(E_1 \cap E_2)$, we can interchange E_1 and E_2 in Eq. 2-9 to obtain

$$P(E_1|E_2) \cdot P(E_2) = P(E_1 \cap E_2) \qquad (2\text{-}10)$$

Equations 2-9 and 2-10 constitute alternative formulations of the *general rule* for multiplication of probabilities.

Let us apply these rules to determining $P(E_2|E_1)$. We can calculate $P(E_1)$, the unconditional or marginal probability that the number of spots is odd by using Eq. 2-6. The simple events which result in odd numbers are e_1, e_3, and e_5. The probability of each of these events occurring is, by Eq. 2-4, 1/6. Consequently, $P(E_1) = 1/2$. The outcomes having a number of spots greater than 3 are e_4, e_5, and e_6. Consequently, $P(E_2) = 1/2$. Since e_5 is common to E_1 and E_2, the $P(E_1 \cap E_2) = 1/6$. To find $P(E_2|E_1)$, we use Eq. 2-8:

$$P(E_2|E_1) = \frac{P(E_1 \cap E_2)}{P(E_1)} = 1/6 \div 1/2 = 1/3$$

Notice that we have *revised* the probability of the occurrence of E_2 (from 1/2 to 1/3) because we were given the additional information that the number of spots is odd.

Independent Events. If events E_1 and E_2 are *independent*, then E_1 does not influence the occurrence of E_2. Consequently, we are unable to reevaluate the number of simple events in an experiment on the basis of additional information. In other words, $P(E_1|E_2) = P(E_1)$ for independent events and our expression for combining probabilities becomes

$$P(E_1 \cap E_2) = P(E_1) \cdot P(E_2) \qquad (2\text{-}11)$$

Equation 2-11 is known as the *special rule* for multiplication, and is used for combining probabilities in the case of independent events. As was true with the addition rules, the multiplication rules can be readily extended for use with more than two simple or composite events.

BAYES' THEOREM

Consideration of expressions 2-9 and 2-10 for computing conditional probabilities indicates that we can compute $P(E_1|E_2)$ if $P(E_2|E_1)$ is known. $P(E_1 \cap E_2) = P(E_2 \cap E_1)$, and from Eqs. 2-9 and 2-10 we have

$$P(E_1 \cap E_2) = P(E_1) \cdot P(E_2|E_1)$$

and

$$P(E_2 \cap E_1) = P(E_2) \cdot P(E_1|E_2)$$

Substituting, we have

$$P(E_1) \cdot P(E_2|E_1) = P(E_2) \cdot P(E_1|E_2)$$

and dividing both sides by $P(E_2)$ yields

$$P(E_1|E_2) = \frac{P(E_1) \cdot P(E_2|E_1)}{P(E_2)} \tag{2-12}$$

Equation 2-12 is a simplified statement of the Bayes' theorem and is useful in computing posterior probabilities (probabilities computed after additional information has been obtained), a cornerstone of Bayesian analysis.

For example, suppose we wished to calculate the probability that the number of spots showing is odd, given that the number showing is greater than three. Symbolically, we want $P(E_1|E_2)$. By Bayes' theorem, this probability is

$$P(E_1|E_2) = \frac{P(E_2|E_1) \cdot P(E_1)}{P(E_2)} = \frac{1/3 \cdot 1/2}{1/2} = 1/3$$

Notice that we have reasoned backward—from effect to cause, to revise the original (or prior) probability $P(E_1)$ of 1/2 to a new (or posterior) probability of 1/3 by using Bayes' theorem to evaluate the additional information that the face which appeared contained more than three spots.

The above statement of Bayes' theorem can be generalized to more than two events. For example, consider an experiment which has mutually exclusive outcomes that can be represented by $E_1, E_2, E_3, \ldots, E_n$, which partition the set E, i.e., $E = E_1 \cup E_2 \cup E_3 \cup \cdots \cup E_n$. Suppose there is a set F of outcomes in which we are interested. A Venn diagram of the sample spaces would appear as in Fig. 2-3.

Now if we know the outcome F has occurred, we can focus on the sample space F and compute the probabilities of occur-

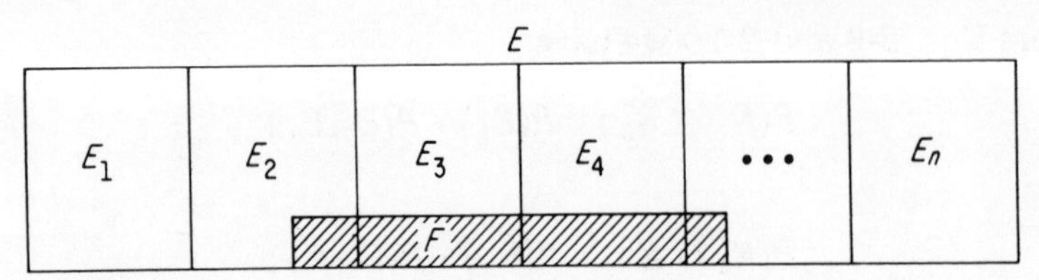

Fig. 2-3. Illustration of generalized application of Bayes' theorem.

rence of each of the sets E_i by Eq. 2-8

$$P(E_i|F) = \frac{P(E_i \cap F)}{P(F)} \qquad \text{for } i = 1, 2, 3, \ldots, n \tag{2-13}$$

Note that we can write

$$P(E_i \cap F) = P(E_i) \cdot P(F|E_i) \tag{2-14}$$

and

$$P(F) = P(E_1) \cdot P(F|E_1) + P(E_2) \cdot P(F|E_2) + \cdots + P(E_n) \cdot P(F|E_n) \tag{2-15}$$

If we substitute Eqs. 2-14 and 2-15 into Eq. 2-13, we get the standard form of Bayes' theorem:

$$P(E_i|F) = \frac{P(E_i) \cdot P(F|E_i)}{\sum_{i=1}^{n} P(E_i) \cdot P(F|E_i)} \qquad \text{for any } i = 1, 2, 3, \ldots, n \tag{2-16}$$

TREE DIAGRAMS

A useful technique for the analysis of problems involving probabilities and logical outcomes is to construct a *tree diagram*. To illustrate this technique, let us consider the following two-stage problem. The first stage is to choose one of two dice. The dice look alike, but only one is a fair die; the other has five spots on four of its faces and two spots on the other two faces. The second stage of the experiment is to toss the die selected and note the number of spots that appear. We wish to calculate the probabilities of occurrence of each of the six possible simple events.

In drawing any tree diagram, three steps are followed. First, the "branches" of the tree are sketched, each branch corresponding to one logical outcome for each stage of the problem. Second, compute the probability of each simple event, and determine the outcome of each event. Enter these numbers on the appropriate branches. Note that the sum of probabilities for any given node must equal one. Third, compute the required information by "reading" the appropriate branches.

For the present problem, there are six outcomes (number of spots is one, two, three, four, five, or six) if the fair die is chosen; by Eq. 2-4 each has a probability of occurrence of 1/6. The "crooked" die has only two outcomes: two spots or five spots. Since two faces have two spots and four faces have five spots, the probabilities are found by Eq. 2-4 to be 1/3 and 2/3, respectively. Since we are not allowed to examine the dice before choosing one, we assume that each die is equally likely to be chosen. The total of the probabilities of the two events must be 1, so the probability of choosing a die is 1/2. The tree diagram for this problem is given in Fig. 2-4.

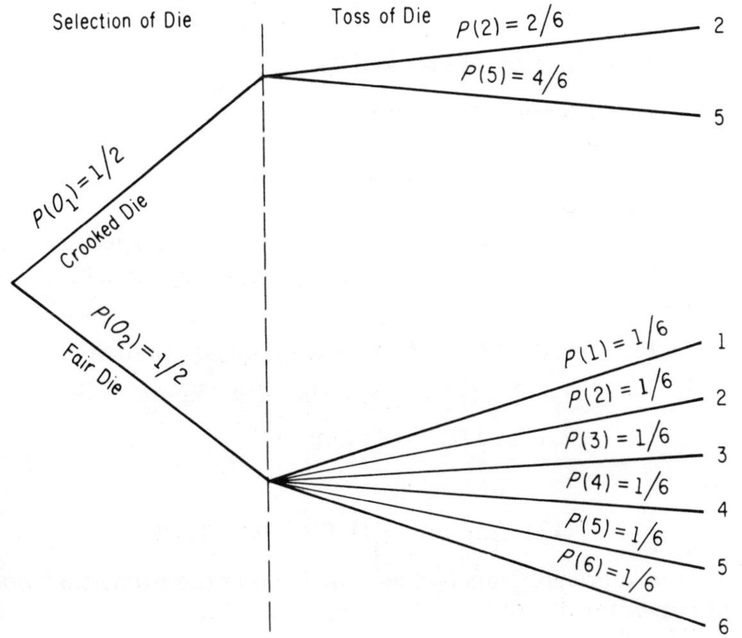

Fig. 2-4. Tree diagram for die selection and toss problem.

Once the tree has been drawn, the third step is to analyze it. Any number of questions can be answered. Suppose, for example, we wish to know the unconditional (or marginal) probability of obtaining five spots. We note that a five can occur in two ways. Following the top branch, corresponding to the event that the crooked die is chosen, the probability of obtaining a five is 4/6. If the fair die is chosen (bottom branch), the probability of obtaining a five is 1/6. Consequently, the unconditional or marginal probability of obtaining a five in this situation is calculated by multiplying the probabilities on each branch and adding them together:

$$(1/2) \cdot (4/6) + (1/2) \cdot (1/6) = 5/12$$

The reader will note that we could have obtained the answer directly by applying Eq. 2-9 and then Eq. 2-6:

$$P(5 \cap O_1) = P(5|O_1) \cdot P(O_1) = 4/6 \cdot 1/2 = 4/12$$

$$P(5 \cap O_2) = P(5|O_2) \cdot P(O_2) = 1/6 \cdot 1/2 = 1/12$$

and

$$P(5) = P(5 \cap O_1) + P(5 \cap O_2) = 1/12 + 4/12 = 5/12$$

where O_1 = selection of crooked die

O_2 = selection of fair die

This result is somewhat more easily visualized when the tree diagram is employed.

The probability of obtaining any number of spots can be obtained simply by tracing the appropriate branch. For example,

$$\begin{aligned} P(2) &= 2/6 \cdot 1/2 \quad \text{from the ``crooked'' branch} \\ &+ 1/6 \cdot 1/2 \quad \text{from the ``fair'' branch} \\ &= 2/12 + 1/12 = 3/12 \end{aligned}$$

Also

$$P(1) = (0 \cdot 1/2) + (1/6 \cdot 1/2) = 1/12$$

since the crooked branch does not include the event that one spot appears.

An alternative method of summarizing the results of such an experiment is to construct a table of joint and marginal probabilities. For the present experiment, we might employ the following symbols:

x_i = event that i spots appear ($i = 1, \ldots, 6$)

O_j = selection of jth die

$j = 1$: crooked die
$j = 2$: fair die

The following relationships are known:

$$P(O_1) = P(O_2) = 1/2$$

i.e., each die has an equal chance of being selected:

$$P(x_i \cap O_j) = P(x_i|O_j) \cdot P(O_j)$$

the joint probability of the appearance of i spots on the jth die is calculated by applying Eqs. 2-9 and 2-10, and

$$P(O_j|x_i) = \frac{P(x_i|O_j) \cdot P(O_j)}{P(x_i)}$$

the probability that the jth die was selected, given that i spots has appeared, is given by Bayes' theorem, Eq. 2-16.

The computations are shown in Table 2-1.

TABLE 2-1

Joint, Marginal, and Revised Probabilities for Die Selection and Toss Problem

x_i	O_1	O_2	$P(x_i)$	$P(O_1 \mid x_i)$	$P(O_2 \mid x_i)$
1	0	1/12	1/12	0	1
2	2/12	1/12	3/12	2/3	1/3
3	0	1/12	1/12	0	1
4	0	1/12	1/12	0	1
5	4/12	1/12	5/12	4/5	1/5
6	0	1/12	1/12	0	1
$P(O_j)$ =	6/12	6/12	12/12		

The first column enumerates the six logical possibilities in the problem. Columns 2 and 3 show the joint probabilities of the occurrence of each x_i and O_j. The column totals are the unconditional or marginal probabilities of each O_j. Each row of column 4 is obtained by summing the corresponding rows i columns 2 and 3. These sums are the marginal probabilities c obtaining each x_i. The total of this column is 1, the tota probability. The total probability can also be obtained by sum ming the marginal probabilities of the occurrence of each O horizontally. Columns 5 and 6 present the revised probabilit of the occurrence of each O_j, given the number of spots appear ing. Note that the sum of the entries in a given row of column 5 and 6 must be 1, since either O_1 or O_2 must occur. Colum totals for these columns are meaningless.

MATHEMATICAL EXPECTATION

We can extend our "crooked-die" problem to illustrat another probability concept of importance in this book. Thi concept is *mathematical expectation*, the average result whic could be expected if an experiment were repeated a large num ber of times. This concept is defined in the following manner.

Let E denote a discrete random variable (set of n dis tinguishable outcomes) which can assume values $e_1, e_2, \ldots$ e_n, with respective probabilities $P(e_1), P(e_2), \ldots, P(e_n)$ for set of mutually exclusive events. Then the mathematical ex pectation of E (denoted by $E[E]$) is given by

$$E[E] = e_1 P_1 + e_2 P_2 + \cdots + e_n P_n = \sum_{i=1}^{n} e_i P_i \qquad (2\text{-}17)$$

where $P_i = P(e_i)$ to simplify the notation. Equation 2-17 say that the mathematical expectation of a set of events is found b multiplying the value of each event in the set by its probability and summing the products. Mathematical expectation can als be defined for continuous random variables using the calculus.

Using Eq. 2-17 the mathematical expectation of the numbe of spots showing after tossing the crooked die is

$$E[O_1] = 2(1/3) + 5(2/3) = 12/3 = 4$$

Of course, four spots would not appear on a given toss of th crooked die, but this result makes sense when we think i terms of averages. If the experiment were repeated a large num ber of times, five spots would occur more often than two spot and the average result would be four spots. The mathematical expectation of the outcome from tossing the fair die is

$$E[O_2] = 1(1/6) + 2(1/6) + 3(1/6) + 4(1/6) + 5(1/6) + 6(1/6)$$
$$= 21/6 = 3.5$$

The mathematical expectation of the two-stage situation (choosing a die and then tossing it) is

$$E[O] = E[O_2] \cdot P(O_2) + E[O_1] \cdot P(O_1)$$
$$= 3.5(1/2) + 4(1/2) = 3.75$$

When the random variable we are studying is expressed in monetary units, as is the case with most business problems, expression 2-17 can be used to compute *Expected Monetary Value* (EMV)—a useful choice criterion as is shown in Chapter 3.

MATRIX ALGEBRA

The reader familiar with matrix notation will recognize that we could define a $1 \times n$ (1-row by n-column) matrix E of outcomes, and an $n \times 1$ (n-row by 1-column) matrix P of probabilities, and obtain Eq. 2-17 by matrix multiplication.

$$E[E] = E \cdot P = [e_1, e_2, \ldots, e_n] \begin{bmatrix} P_1 \\ P_2 \\ \vdots \\ P_n \end{bmatrix} = \sum_{i=1}^{n} e_i P_i$$

In general, if A is an m-row $\times$ n-column matrix, and B is an n-row $\times$ p-column matrix, the m-row $\times$ p-column matrix $C = AB$ is defined by

$$c_{ij} = \sum_{k=1}^{n} a_{ik} \cdot b_{kj} \qquad (2\text{-}18)$$

where c_{ij} = the element in the ith row and jth column of C

a_{ik} = the element in the ith row and kth column of A

b_{kj} = the element in the kth row and the jth column of B

For example, if A is a 2-row $\times$ 3-column matrix, denoted $[A_{ik}]_{2,3}$ and B is a 3-row $\times$ 2-column matrix, denoted $[B_{kj}]_{3,2}$, then $C = AB$ is a 2-row × 2-column matrix $[C_{ij}]_{2,2}$, defined as

$$\begin{bmatrix} a_{11} & a_{12} & a_{13} \\ a_{21} & a_{22} & a_{23} \end{bmatrix} \begin{bmatrix} b_{11} & b_{12} \\ b_{21} & b_{22} \\ b_{31} & b_{32} \end{bmatrix}$$

$$= \begin{bmatrix} a_{11}b_{11} + a_{12}b_{21} + a_{13}b_{31}, & a_{11}b_{12} + a_{12}b_{22} + a_{13}b_{32} \\ a_{21}b_{11} + a_{22}b_{21} + a_{23}b_{31}, & a_{21}b_{12} + a_{22}b_{22} + a_{23}b_{32} \end{bmatrix}$$

For example,

$$\begin{bmatrix} 1 & 2 & 4 \\ 3 & 7 & 5 \end{bmatrix} \begin{bmatrix} 1 & 4 \\ 3 & 6 \\ 2 & 0 \end{bmatrix} = \begin{bmatrix} 1 \cdot 1 + 2 \cdot 3 + 4 \cdot 2, & 1 \cdot 4 + 2 \cdot 6 + 4 \cdot 0 \\ 3 \cdot 1 + 7 \cdot 3 + 5 \cdot 2, & 3 \cdot 4 + 7 \cdot 6 + 5 \cdot 0 \end{bmatrix}$$

$$= \begin{bmatrix} 15 & 16 \\ 34 & 54 \end{bmatrix}$$

Matrix algebra is an efficient method of computation for data in tabular form. Since much of the data in the following chapters is presented in tables, we shall use matrix algebra to expedite our computations.

SUMMARY

Bayesian analysis treats uncertainty by enumerating possible states of nature, assigning probabilities of occurrence to these states, and calculating expected values for each state. Consequently the rudiments of probability theory must be understood by anyone wishing to apply Bayesian procedures to decision problems. In this chapter we have discussed the objective (relative frequency) and subjective (degree of belief) meanings of probability. The remainder of the chapter was devoted to a review of elementary probability concepts. From the three axioms for assigning probabilities to events in a set were derived rules for addition and multiplication of probabilities for mutually exclusive, independent, and dependent events. We then derived Bayes' theorem for the revision of probabilities. Next, the usefulness of tree diagrams and tables of joint and marginal probabilities was illustrated, and the concept of mathematical expectation was explained. Finally, the use of matrix algebra as a computational tool was reviewed. The following chapters apply these concepts to problems in marketing.

REFERENCES

Feller, William. *An Introduction to Probability Theory and its Applications.* 2 vols. 3d ed. New York: John Wiley & Sons, Inc., 1968.

This book is a standard reference in the field, but is not easy reading for the mathematically unsophisticated.

Fishburn, Peter C. *Decision and Value Theory.* New York: John Wiley & Sons, Inc., 1964.

Comprehensive treatment of decision theory and the role of probability theory in studies of the decision process.

Kemeny, John G. et al. *Finite Mathematics with Business Applications.* Englewood Cliffs, N.J.: Prentice-Hall, Inc., 1962.

Very readable elementary treatment of topics reviewed in this chapter.

PROBLEMS

2-1. Given $P(A) = .25, P(B) = .30$ and $P(A \cap B) = .15$; find

(a) $P(A')$ (d) $P(A \cup B)$

(b) $P(B')$ (e) $P(A' \cap B')$

(c) $P(A \cap B')$ (f) $P(A' \cup B')$.

2-2. Given that a pair of fair dice is thrown one time, find $P(X)$, (X = sum of spots on faces of dice), for the following values of X:

(a) $X \neq 7$ (c) $X < 10$

(b) $X \leqslant 4$ (d) $4 < X \leqslant 9$

2-3. Of a group of 100 students, 40 are philosophy majors, 30 are mathematics majors and the remainder are budding economists. Ten of the mathematics students have A averages. The economists are equally divided among grades A, B, and C. There are 40 C students in total; 18 of these are philosophy majors. Six philosophy majors are B students. Using P, M, E to designate majors and A, B, and C to designate grade average, find

(a) $P(C')$ (d) $P(M|C)$

(b) $P(M \cap C)$ (e) $P(C|M)$

(c) $P(M \cup C)$ (f) $P(M|A')$

2-4. Three salesmen cover a particular territory. Salesman A writes 1/2 of the total orders, B writes 1/3 of the orders and C writes the remainder. The orders are complex, so the salesmen sometimes make errors. Past records indicate that salesman A erroneously records 1/3 of the orders he writes; B errors 10 percent of the time and C commits errors on 1/5 of his orders.

(a) Show that the probability that an error occurs on a particular order is 7/30 by constructing a table of joint and marginal probabilities, and by drawing a tree diagram. (Hint: use a common denominator of 30.)

(b) If an error were committed on a given order, what is the probability that A committed the error?

(c) Given that the revenue from a sale is \$60 and that an error adjustment costs \$90, determine which of the three salesman is best.

(d) Show that the expected revenue from a sale in this territory is \$39.

CHAPTER 3

Decision Theory: An Introduction

This chapter presents fundamentals of decision theory. The basic decision model and terminology are introduced, and various decision criteria are illustrated and critiqued. The purpose of this chapter is to review various approaches to decision making and to demonstrate the efficacy of the expected monetary value criterion which is commonly used in Bayesian analysis.

THE BASIC DECISION MODEL

The situation any decision maker faces can be described as a choice among at least two courses of action that are functionally related to certain environmental conditions, called *states of nature*, which are beyond the decision maker's control. These conditions affect the courses of action in such a way that the interaction produces a unique result. This situation can be symbolized as follows.

$$V_{ij} = f(A_i, S_j) \qquad (3\text{-}1)$$

where A_i = the ith course of action available to the decision maker ($i = 1, 2, \ldots, m$) possible courses of action

S_j = jth state of nature that can occur ($j = 1, 2, \ldots, n$) possible states

V_{ij} = the value of the result of the interaction of the ith course of action and the jth state of nature

$f(\cdot)$ = functional relationship between the independent variables A_i and S_j and the dependent variable V_{ij}

This general model can be used to formulate any type of decision problem. There are, however, two difficulties associated with using the model. First, each V_{ij} must be defined precisely—that is, it must be quantified. To accomplish this requires definition of a measure of value, and specification of the functional relationship between each V_{ij} and the interaction of each A_i and S_j.

Economic theory postulates the ideal measure of value: utility, the amount of satisfaction resulting from a specific V_{ij}. But there is no generally agreed-upon empirical standard for measuring utility. Interpersonal and temporal comparisons of utility scales are particularly difficult. Fortunately, many of the problems which marketing management faces can be resolved in terms of a standard which does have rather wide acceptance: dollars and cents. It is possible to calculate the monetary payoff associated with a specific V_{ij}. In this book, therefore, V_{ij} will be measured in terms of monetary values.[1] This criterion is simply the accounting model: profit = revenue - cost.

A second difficulty inherent in the basic decision model is that after the V_{ij}'s are determined, a criterion must be used to determine the preferred A_i. Consequently, a criterion model must be established to guide the decision maker's choice among the available courses of action. The following section reviews suggested criteria which can be applied to various types of decision environments which the manager might face.

TYPES OF DECISION ENVIRONMENTS

To make our discussion a little more concrete, let us suppose that a marketing manager faces the following decision situation. He must decide whether to expand his franchised ice-cream parlor, or to continue his present operation. Let the decision not to expand be denoted by A_1, the decision to expand by A_2. Assume that the possible states of nature which will exist when the decision is made consist of only two: the market in which the franchise is located has low sales potential, S_1, or it has high

[1] We shall modify this statement, at least conceptually, when we discuss Expected Utility Value in Chapter 7.

sales potential, S_2. In terms of the basic decision model, this situation is described in Table 3-1.

TABLE 3-1

The Basic Decision Model

Course of Action, A_i	State of Nature, S_j	
	S_1 Low Sales Potential	S_2 High Sales Potential
A_1 : Retain present facilities.	$V_{11} = f(A_1, S_1)$	$V_{12} = f(A_1, S_2)$
A_2 : Expand facilities	$V_{21} = f(A_2, S_1)$	$V_{22} = f(A_2, S_2)$

If sales potential is high, expanded facilities will generate more revenue than will present facilities. But if sales potential is low, the additional revenue will not cover the expansion costs. To evaluate this model each V_{ij} must be quantified. Since we have decided to use monetary value as a unit of measurement, the next step is to assemble revenue and cost data, and calculate the payoff associated with each V_{ij}. The data-collection processes present a number of interesting problems, but these are problems related more to accounting and marketing research (as traditionally defined) than to our topic, the decision-making process.[2] For this reason we shall make the rather handy assumption that the data are available. The following information results:

1. Expansion entails some cost. The manager would like to recover this cost within a specified time period (planning horizon).
2. If sales potential is high, the present value of the net payoff will be \$9,000 for an expanded facility and \$6,000 for the present facility.[3]

[2] The interested reader can pursue these topics in such books as C. H. Sevin, *Marketing Productivity Analysis* (New York: McGraw-Hill Book Company, 1965); H. W. Boyd, Jr. and Ralph Westfall, *Marketing Research: Text and Cases*, 2d ed. (Homewood, Ill.: Richard D. Irwin, Inc., 1964), R. N. Anthony, *Management Accounting*, 3d ed. (Homewood, Ill.: Richard D. Irwin, Inc., 1964). A forthcoming book, Keith K. Cox and Ben M. Enis, *The Marketing Research Process; Purchasing Relevant Information for Decision Making* (Pacific Palisades, Calif.: Goodyear Publishing Company), treats these problems within the context of the decision process presented in this book.

[3] The present value concept, i.e., recognizing the time value of money, is an integral aspect of managerial accounting For an explanation, see Anthony, *ibid.*, Chap. 18.

3. If sales potential is low, the present value of the net payoff will be $2,000 for an expanded facility and $3,000 for the present facility.

The payoff matrix derived from these facts is shown in Table 3-2.

Given this payoff matrix, the manager must choose either A_1 or A_2. The nature of his knowledgc of the environment significantly influences the criterion used to make this choice. This environmental knowledge may be one of three types—certainty, risk, or uncertainty.[4]

TABLE 3-2

Franchise Expansion Decision-Model Data (thousands of dollars)

Course of Action, A_i	State of Nature, S_j	
	S_1 Low Sales Potential	S_2 High Sales Potential
A_1: Retain present facilities . .	$V_{11} = 3$	$V_{12} = 6$
A_2: Expand facilities	$V_{21} = 2$	$V_{22} = 9$

Certainty means that the decision maker knows which state of nature will occur. In our example, certain knowledge means that the manager knows whether S_1 (low sales potential) or S_2 (high sales potential) is true. If his objective is the usual one of maximizing the payoff, he will make the decision which results in the higher V_{ij} for the S_j which actually will occur. If S_1 is to occur, he will choose A_1, since V_{11} = $3,000 is greater than V_{21} = $2,000. Conversely, if he knows that S_2 is the true state of nature, he will select A_2, since V_{22} = $9,000 is greater than V_{12} = $6,000.

The marketing manager does not often have certain knowledge of the state of nature which will occur at some time in the future. He usually must predict which state of nature will occur. If he knows the probability of occurrence of each of the various states of nature, he can use expected values of the A_i's to determine which decision to make. This situation is referred to as *decision making under risk*. For example, if historical data were available which indicated that the relative frequency of

[4] The classic conceptual work in this area is F. H. Knight, *Risk, Uncertainty, and Profit* (Boston: Houghton Mifflin, Company, 1921).

occurrence was .4 for S_1 and .6 for S_2, the manager might reason as follows:

$$\text{EMV} = \begin{bmatrix} 3 & 6 \\ 2 & 9 \end{bmatrix} \begin{bmatrix} .4 \\ .6 \end{bmatrix} = \begin{bmatrix} 1.2 + 3.6 \\ .8 + 5.4 \end{bmatrix} = \begin{bmatrix} 4.8 \\ 6.2 \end{bmatrix}$$

Since the EMV $[A_2]$ = \$6,200 is greater than the EMV $[A_1]$ = \$4,800, the manager would choose A_2.

Decision making under risk requires that the manager know (or be able to reliably estimate) the true probability of occurrence of each S_j. That is, either he must know the parameters of the process which generates the probabilities, or have long-run frequency data on past occurrences. For example, an actuary uses long-run frequency data on deaths to determine the probability that an individual, of a given age, with a particular ailment, profession, etc., will live to be a specific age, and therefore assess an appropriate life insurance premium.

Unfortunately, events in marketing are often unprecedented. A particular new product, for example, can be introduced only once. A certain salesman can be added to our staff now, but the opportunity to hire him may not recur. Since long-run frequency data cannot be generated for most marketing situations, the marketing manager does not even know the probability of occurrence of each S_j. Decision making in this situation is known as decision making under *uncertainty*. The decision criterion is of particular importance in this situation. A number of such criteria have been suggested by decision theorists, which indicates the complexity of this type of decision problem. Six of the more generally suggested criteria are discussed below.

CHOICE CRITERIA FOR DECISIONS UNDER UNCERTAINTY

Generally, one of six criteria is used to guide the decision maker's evaluation of alternative courses of action under uncertainty. Three of these criteria, *maximax*, *maximin*, and *minimax regret* are game-theory approaches. That is, they treat nature as an opponent to be defeated by strategy. The other three criteria, *the Laplace criterion*, *Expected Monetary Value*, and *Expected Opportunity Loss*, are probabilistic approaches to decision making. Each of these criteria is applied to the data summarized in Table 3-2.

Maximax Criterion

Suppose our decision maker is optimistic. He believes that the most favorable state of nature will occur in a given situation. Consequently, he will act so as to maximize his maximum payoff. Given that criterion and the situation summarized in Table 3-2, the decision maker will choose A_2 (expand facilities), because the payoff will be $9,000 if S_2 (high sales potential) is the true state of nature. Of course, S_2 is the better (potentially more profitable,) state of nature. A decision criterion which assumes that the best possible result will invariably occur—that nature will always act benevolently, is perhaps not too realistic a policy in all cases. In certain situations, however, it may be the only feasible criterion to use. If for example the firm would be bankrupt unless at least a $9,000 profit were generated by the end of the planning period, the decision maker might use the maximax criterion in desperation rather than from optimism. Column 3 of Table 3-3 summarizes this analysis.

TABLE 3-3
Franchise-Expansion Problem—Analysis by Maximax and Maximin
(thousands of dollars)

Course of Action, A_i	State of Nature, S_j		Criterion	
	S_1 Low Sales Potential	S_2 High Sales Potential	Maximax	Maximin
A_1 : Retain present facilities	3	6	6	3 (max)
A_2 : Expand facilities	2	9	9 (max)	2

Maximin Criterion

Maximin is the criterion of pessimism. The decision maker always believes that the worst event possible in a particular situation is the one which will occur. Given this belief, the rational decision rule is to maximize the minimum payoff. From Table 3-2 we see that, for A_1, the payoff for S_1 is smaller than for S_2. Consequently, the minimum payoff for A_1

(\$3,000) results from the occurrence of S_2. Similarly, the minimum payoff from A_2 is \$2,000. Since \$3,000 > \$2,000, the decision maker using the maximin criterion would choose A_1. He would thus be assured of earning at least \$3,000. Column 4 of Table 3-3 summarizes this analysis.

The maximin criterion assumes that nature is essentially malevolent, and that the decision maker will act conservatively. Moreover, this criteron neglects information which might be relevant. If V_{22} were \$9,000,000 (and all other figures remained as in Table 3-3), the maximin criterion would still indicate A_1 as the preferred decision. The very high potential payoff associated with A_2 would be ignored.

Minimax Regret Criterion

Another essentially pessimistic decision criterion would be to minimize the maximum loss, or regret, which could have resulted from a given decision. This analysis employs "hindsight" to construct a loss matrix as follows. Had S_1 occurred, the decision maker should have chosen A_1, since by so doing he would have earned \$3,000 rather than \$2,000. A_1 was the best decision for the situation which occurred. Had he chosen A_2, he would have "regretted" it, because he would have forgone the opportunity to earn \$3,000 - \$2,000 = \$1,000. Similarly, had S_2 occurred, he should have selected A_2. Since A_2 was the best decision for the situation which actually occurred, there is no opportunity loss associated with A_2. Had he chosen A_1, however, his opportunity loss, or regret, would have been \$9,000 - \$6,000 = \$3,000. To formalize this reasoning, let L_{ij} be the opportunity loss associated with the ith course of action and the jth state of nature. Specifically, $L_{ij} = V_j^* - V_{ij}$ where $V_j^* = \max V_{ij}$ for a given j. Given this information, we can now construct Table 3-4 to summarize the discussion so far.

Now, if our decision maker chooses A_1, his maximum regret will be \$3,000. If he chooses A_2, the maximum regret is \$1,000. He should minimize his maximum regret, therefore, by choosing A_2.

The fact that the maximin and minimax-regret criteria indicate opposite conclusions merits comment. Since both are pessimistically based criteria, one might expect them to produce similar results. In this case, however, maximin indicated that

TABLE 3-4
Franchise-Expansion Problem—Analysis by Minimax Regret
(thousands of dollars)

Course of Action, A_i	State of Nature, S_j		Maximum Regret
	S_1 Low Sales Potential	S_2 High Sales Potential	
A_1 : Retain present facilities .	$L_{11} = 0$	$L_{12} = 3$	3
A_2 : Expand facilities	$L_{21} = 1$	$L_{22} = 0$	1 (min)

the minimum payoff will be maximized by retaining the present facility; minimax-regret analysis showed that opportunity losses will be minimized by expanding facilities. The apparent discrepancy is a reflection of the fact that maximin does not take the magnitude of favorable results into account. Minimax regret, on the other hand, uses the difference between the payoff of the most favorable outcome and the other payoffs to determine the best decision. In the present case, the difference in payoff between expanding and not expanding would result in a substantial opportunity loss from not expanding when sales potential is high. This fact is ignored by the maximin criterion.

Note again the essential pessimism of the minimax-regret criterion: nature is always expected to do her worst. Moreover, like maximax and maximin, the minimax-regret criterion assumes that one state of nature will occur. No consideration is given to the likelihood of occurrence of the expected alternative. Consideration of these likelihoods adds a dimension to the decision problem.

Laplace Criterion

The Laplace criterion recognizes that in the face of uncertainty, probabilities could be employed in predicting the occurrence of future events. In the absence of exact knowledge of the probabilities, the Laplace criterion states that each outcome should be assigned equal probabilities. If this procedure were applied to the present case, each of the two events S_1 and S_2 would be assigned a probability of .5. Then the EMV $[A_i]$ would be:

$$\text{EMV}\,[A_i] = \begin{bmatrix} 3 & 6 \\ 2 & 9 \end{bmatrix} \begin{bmatrix} .5 \\ .5 \end{bmatrix} = \begin{bmatrix} 1.5 + 3.0 \\ 1.0 + 4.5 \end{bmatrix} = \begin{bmatrix} 4.5 \\ 5.5 \end{bmatrix}$$

Thus the decision maker would maximize his expected payoff by choosing A_2.

The Laplace criterion has its uses. In tossing a fair die, most people would be willing to assign a probability of 1/6 to the event that a given number of spots appears on a particular toss. However, if a bettor consistently applied this criterion to horse racing, he might lose his shirt: in most races, all horses do not have an equal chance of winning.[5] Most marketing decisions resemble horse racing rather than die tossing—they are usually concerned with events which are not equally likely to occur. Consequently, the marketing manager should look for a better method of assigning probabilities to outcomes.

Expected Monetary Value

The concept of expected value, as reviewed in Chapter 2, has been introduced in our earlier discussion of decision making under risk, and in using the Laplace criterion. The usefulness of this concept as an approach to decision making has been demonstrated by the earlier discussions. Knowing the EMV of various courses of action can be very helpful to the manager in making choices.

Under uncertainty, however, the decision maker does not know the probabilities of occurrence of the possible states of nature, so the traditional concept of expected value cannot be applied. But in most decision making situations in marketing, the manager has some idea, some intuitive "feel," for the probabilities of occurrence of the various states of nature. This subjective assessment is usually a function of the manager's knowledge and general experience.

The Bayesian approach to decision making argues that these "personalistic probabilities" should not be ignored. Indeed, they are not ignored by a manager in an actual decision situation. He evaluates the factors of the decision in his head, implicitly assigning probabilities to the occurrence of those factors about which he is uncertain. The Bayesian approach simply formalizes (makes explicit) the decision-making process which most managers actually follow.

Suppose that based on his experience in estimating sales

[5] This example is amplified in delightful fashion in I. D. J. Bross, *Design for Decision* (New York: The Macmillan Company, 1953), Chap. 4.

potential, our manager assigned a probability of .3 to S_1 (low sales potential) and .7 to S_2. Consequently, the EMV of A_1 (retain present facilities) and A_2 (expand facilities) are given by

$$\text{EMV}[A_i] = \begin{bmatrix} 3 & 6 \\ 2 & 9 \end{bmatrix} \begin{bmatrix} .3 \\ .7 \end{bmatrix} = \begin{bmatrix} .9 + 4.2 \\ .6 + 6.3 \end{bmatrix} = \begin{bmatrix} 5.1 \\ 6.9 \end{bmatrix}$$

The manager wants to maximize the expected payoff of his decision, so he should choose A_2, since $\text{EMV}[A_2] = \$6{,}900$ is greater than $\text{EMV}[A_1] = \$5{,}100$.

Notice that had the manager assigned a different probability, say, .8, to S_1, his results would be different.

$$\text{EMV}[A_i] = \begin{bmatrix} 3 & 6 \\ 2 & 9 \end{bmatrix} \begin{bmatrix} .8 \\ .2 \end{bmatrix} = \begin{bmatrix} 2.4 + 1.2 \\ 1.6 + 1.8 \end{bmatrix} = \begin{bmatrix} 3.6 \\ 3.4 \end{bmatrix}$$

He should choose A_1 for this probability of occurrence of S_1. This calculation indicates that, as the probability that S_1 (low sales potential) is the true state of nature increases, the conservative alternative, A_1, becomes more attractive, relative to A_2. Conversely, as the probability that S_2 (high sales potential) is true increases, alternative A_2 becomes relatively more valuable. This reasoning suggests that the manager could determine an indifference point, a set of probabilities of the occurrence of the two states of nature which equalizes the payoff of the two courses of action. Figure 3-1 illustrates this procedure. The payoffs of the two courses of action are plotted as a function of the probability that S_2 (high sales potential) occurs.[6]

The figure shows that the indifference probability for E_2 is .25. By Axiom 2 of probability theory, the corresponding probability for E_1 is $1.00 - .25 = .75$. Of course, Fig. 3-1 could have been plotted as a function of E_1, in which case the .75 probability would be read from the figure and .25 obtained by Axiom 2. An alternative method of calculating the indifference point is to note that, at this point, the expected monetary value of the two decisions is equal.

[6] Assuming that each payoff function is linear, both can be plotted by calculating conditional payoffs for any two probabilities, and connecting the points with a straight line.

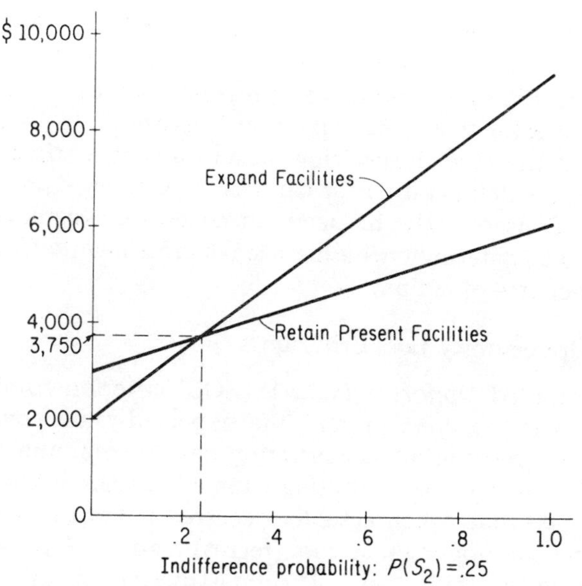

Fig. 3-1. Illustration of indifference probability for franchise-expansion problem.

$$\text{EMV}[A_1] = \text{EMV}[A_2]$$

Letting p be the indifference probability for the occurrence of E_1, we solve for p:

$$V_{11}p + V_{12}(1 - p) = V_{21}p + V_{22}(1 - p)$$
$$3p + 6(1 - p) = 2p + 9(1 - p)$$
$$4p = 3$$
$$p = 3/4 = .75$$

Check:

$$\text{EMV}[A_1] = \text{EMV}[A_2]$$
$$3(.75) + 6(.25) = 2(.75) + 9(.25)$$
$$2.25 + 1.50 = 1.50 + 2.25$$
$$3.75 = 3.75$$

This concept provides useful information for the decision maker in that it indicates the "sensitivity" of the decision to his probability estimates. The calculations indicate that, for any probability of high sales potential less than .25, it is better (in terms of EMV) to retain the present facilities. For any probability greater than .25, expansion is expected to be more profitable. In this case, the decision maker assigned a probability of .70 to S_2, which is rather far from the indifference probability of .25. Consequently, he need not review his assignment of probabilities to determine whether a small error might affect his choice of a course of action.

Expected Opportunity Loss Criterion

The *Expected Opportunity Loss* (EOL) criterion combines the minimax-regret concept and the expected-value concept. The decision maker using this criterion desires to minimize his opportunity loss (the loss resulting from a less than optimal decision for the event which actually occurred.) For the reader's convenience, the opportunity loss (regret) matrix of Table 3-4 is reproduced as Table 3-5. The construction of this table should be reviewed if necessary.

TABLE 3-5

Franchise-Expansion Problem—Opportunity-Loss Matrix
(thousands of dollars)

Course of Action, A_i	State of Nature, S_j	
	S_1 High Sales Potential	S_2 Low Sales Potential
A_1: Retain present facilities .	0	3
A_2: Expand facilities	1	0

Since the manager assigned the probabilities .3 and .7 to S_1 and S_2 respectively, the EOL would be

$$\text{EOL}[A_i] = \begin{bmatrix} 0 & 3 \\ 1 & 0 \end{bmatrix} \cdot \begin{bmatrix} .3 \\ .7 \end{bmatrix} = \begin{bmatrix} 0 & +\ 2.1 \\ .3 + 0 & \end{bmatrix} = \begin{bmatrix} 2.1 \\ .3 \end{bmatrix}$$

Since the decision maker desires to minimize his EOL, he should choose A_2. Note that the EOL criterion and the EMV criterion both indicated that A_2 was the best course of action. This is not a coincidence. Both criteria will always indicate the same decision. Specifically, the difference between the EMV of the optimum decision and the EMV of any other decision is equal to the difference between the EOL of the other decision and the EOL of the optimum decision. This can be symbolized as follows.

$$\text{Max EMV}[A_i] - \text{EMV}[A_i] = \text{EOL}[A_i] - \text{Min EOL}[A_i]$$

for any i. In terms of the present problem,

$$\$6{,}900 - \$5{,}100 = \$2{,}100 - \$300 = \$1{,}800$$

The relationship between these criteria is developed more fully in Chapter 4.

Usefulness of the Criteria Discussed

Each of the six criteria discussed may be useful in evaluating decisions under conditions of uncertainty. When the decision maker has reason to be completely optimistic or completely pessimistic and has no other basis for deciding, the game-theory approaches may be more appropriate. However, these criteria represent extremes of expectations concerning outcomes, and also neglect information concerning the magnitude of potential payoffs.

In most cases in evaluating decision making under uncertainty, some likelihood that each of the possible states of nature will occur is available. If these likelihoods can be formulated according to the axioms of probability, EMV will indicate the best course of action in a given case. Since equally likely states of nature are most unusual in marketing, the Laplace criterion is not particularly useful. Moreover, the decision maker usually knows something about the situation in question. If his knowledge, past experience, and intuition can be used to assign probabilities to states of nature, EMV can be employed as a criterion of choice.

The Bayesian view is that quantification of experience is both realistic and desirable. Formal incorporation of personalistic probabilities allows the decision maker to make use of all the information relevant to a given situation, in a manner which checks for completeness and consistency of reasoning and which can be communicated to others. For this reason, EMV (or its converse EOL) is used as the decision criterion in the next three chapters of this book. We shall defer to the last chapter a discussion of other situations for which these two criteria are not appropriate.

DOMINANT AND DOMINATED COURSES OF ACTION

There are two "special cases" of decision situations which should be mentioned at this point. These cases are "dominant" and "dominated" courses of action. A dominant course of action is one which has a payoff as high or higher than any other alternative for all states of nature. In terms of symbols, A_k, is a dominant course of action if

$$V_j^* = V_{kj}$$

for every value of j. Table 3-6 presents an example of a dominant course of action. A_2 has the highest payoff for all states of nature. Consequently, the rational decision maker—the one whose objective is to attain the maximum payoff from the decision situation, would choose A_2. No further analysis would be necessary.

TABLE 3-6

Illustration of a Dominant Alternative

Course of Action, A_i	State of Nature, S_j			
	S_1	S_2	S_3	S_4
A_1	1	5	3	7
A_2	10*	12*	15*	14*
A_3	10	6	4	1
A_4	5	7	9	8

A dominated course of action, on the other hand, is one that has a payoff as low or lower than another alternative for every state of nature. In terms of symbols, alternative A_p is dominated by A_q if

$$V_{pj} \leqslant V_{qj}$$

for every value of j. Table 3-7 presents an illustration of a dominated course of action.

In Table 3-7, A_3 is dominated by A_4. That is, under no state of nature would A_3 constitute a more rational decision than would A_4; in fact, A_3 is an inferior decision (lower payoff) for all states except S_3. Consequently, the decision maker would

TABLE 3-7

Illustration of a Dominated Course of Action

Course of Action, A_i	State of Nature, S_j			
	S_1	S_2	S_3	S_4
A_1	8	6	4	7
A_2	3	2	7	9
A_3	1	5	9	7
A_4	5	7	9	8

simply ignore A_3. His feasible courses of action would be A_1, A_2, and A_4. The choice of one of these would then be made according to one of the criteria discussed above.

SUMMARY

This chapter has presented a basic framework for decision making and reviewed several decision criteria. The EMV approach, for many marketing problems, is the most useful criterion in terms of amount of information utilized. This approach and its converse, EOL, form the basis for Bayesian analysis for marketing decisions. The next chapter presents the fundamentals of Bayesian analysis as applied to marketing problems.

REFERENCES

Bross, I. D. J. *Design for Decision*. New York: The Macmillan Company, 1953.

Delightfully written nontechnical exposition of decision theory.

Luce, R. Duncan, and Howard Raiffa, *Games and Decisions*. New York: John Wiley & Sons, Inc., 1957.

The standard reference in the field; not easy reading, but worth the effort.

von Neumann, John, and Oskar Morgenstern. *Theory of Games and Economic Behavior*. Princeton, N.J.: Princeton University Press, 1947.

The pioneering work in the field of decision theory; requires a high level of mathematical sophistication.

PROBLEMS

3-1. As marketing manager for a vending machine company, you must decide whether to stock coffee or beer in your machines at a football stadium on a certain day. You may not stock both coffee and beer on the same day. Of course, coffee sells better than beer on a cold day, and vice versa. Specifically on a cold day, you will make a profit of $4,000 if you stock coffee but you will lose $2,000 if you stock beer. On a hot day, your coffee profit will be $2,000; beer profit will be $12,000. It is your opinion that the probability of a hot day is 40 percent.

(a) Decide whether to stock coffee or beer, using the following criteria:
1. Maximax
2. Maximin
3. Minimax regret
4. Laplace
5. Expected monetary value
6. Expected opportunity loss

(b) At some probability of the occurrence of a cold day, it will not matter whether you decide to stock coffee or beer (assuming EMV as the criterion of choice). Determine this probability.

(c) What decision would you make? Why?

3-2. The following V_{ij} and $P(S_j)$ are given:

Courses of Action, A_i	States of Nature, S_j			
	S_1	S_2	S_3	S_4
A_1	2	3	2	5
A_2	1	4	7	1
A_3	2	6	5	2
A_4	3	5	3	4
$P(S_j)$	.4	.3	.2	.1

(a) Determine the best decision using the following criteria:
1. Maximax
2. Maximin
3. Minimax regret
4. Laplace
5. EMV
6. EOL

(b) Comment on your results from a decision making point of view.

3-3. A baker wishes to know how many of a certain type of cake he should bake. The ingredients for each cake cost $2.50, and he must pay an additional \$.20 for boxing each cake sold to the public. The cakes sell for \$3.70 each. Cakes unsold at the end of the day can be sold to a restaurant for \$.50 each.[7]

(a) Construct a conditional payoff matrix and a conditional loss matrix for 0–5 cakes.

(b) Decide, using maximax, maximin, and minimax regret, how many cakes should be baked.

(c) Given the following demand schedule,

Demand, S_j	Probability, $P(S_j)$
0	.05
1	.10
2	.25
3	.30
4	.20
5	.10

how many cakes should be baked?

[7]Adapted from R. Schlaifer, *Probability and Statistics for Business Decisions* (New York: McGraw-Hill Book Company, 1959), pp. 67-68.

CHAPTER 4

Fundamentals of Bayesian Analysis

This chapter constitutes the heart of the book. The basic elements of Bayesian analysis are presented. The franchise-expansion problem introduced in Chapter 3 will continue to provide a concrete illustration for the concepts presented. The problem is enlarged to include three levels of potential demand (states of nature), and three courses of action. First, the construction of conditional payoff and loss matrices is discussed. The decision maker's prior probability distribution is introduced next. Finally, the value-of-additional-information concept is explained.

CONDITIONAL PAYOFF AND LOSS MATRICES

As we indicated in Chapter 3, the decision maker must be able to quantify the payoff associated with the intersection of each state of nature and each course of action. Symbolically, this process is represented by

$$V_{ij} = f(A_i, S_j)$$

where A_i = ith course of action
S_j = jth state of nature
$f(\cdot)$ = function which maps each A_i and S_j (the independent variables) on to each V_{ij} (the dependent variable), and

V_{ij} = dollar payoff associated with the *i*th course of action *j*th state of nature

The accounting model of the firm

$$\text{Profit} = \text{Revenue} - \text{Costs}$$

provides a means of operationalizing the decision model. The costs, both fixed and variable, of supplying a product are under the control of the decision maker. He determines the quality and quantity of product offered to the market. Revenue is price per unit times number of units sold. Thus, at a given price, revenue is a function of demand, i.e., the existing environment condition or "true state of nature." If, however, the quantity demanded is greater than that supplied, revenue is determined by supply.

In terms of payoffs then, a specific V_{ij} is determined jointly by the *i*th course of action (quantity supplied) and the *j*th state of nature. The decision maker, by choosing a course of action, controls one factor. The other factor, the state of nature which occurs, is beyond his control, so he assigns probabilities to the occurrence of each of the possible states. Multiplying the conditional payoffs V_{ij} by the probabilities and summing yields an expected payoff (EMV) for each course of action.

We can fit this decision model to our franchise-expansion problem. Our decision maker will consider three courses of action: A_1, retain present facilities; A_2, expand facilities; A_3, relocate and expand facilities. Expansion involves some incremental costs, and relocation would entail a greater amount of incremental expense. Three states of nature are considered: market potential has a present value of \$5,000, denoted by S_1; \$15,000, S_2; and \$30,000, S_3. If S_1 is true—if demand for our product will result in revenue from this market having a present value of only \$5,000—our present facilities are just adequate for the share we can obtain; expansion at the present location will not be worthwhile, and relocation would result in a fairly substantial loss. If the present value of market potential is \$15,000, present facilities would be taxed beyond capacity and could not adequately meet the need. Many sales would be lost, although some additional profit would accrue. Expansion of facilities would meet the needs of a market having a present value of \$15,000, and relocation would show a slight profit. If potential has a present value of $30,000, expanded facilities could not handle the demand, although some additional profit would accrue. Relocation would be very profitable in this case.

Let us again make the heroic assumption that sufficient cost and revenue data are available to quantify the above description. The net present value of each of the outcomes are presented in Table 4-1.

TABLE 4-1

Franchise-Expansion Problem—Conditional Payoff Matrix (dollars)

Course of Action, A_i	State of Nature, S_j		
	S_1	S_2	S_3
A_1: Retain present facilities	1,000	1,100	1,200
A_2: Expand facilities	−500	2,000	2,200
A_3: Relocate and expand facilities. .	−2,000	500	3,000

Thus, for example, we see that V_{23}, the payoff from the second action (expand facilities), given that S_3 (potential demand is $30,000 per month) is the true state of nature, is $2,200. The reader will note that we could increase the precision of our analysis by evaluating a larger number of states of nature. This would increase the computation required, but would not alter the concept. We saw no need to complicate our hypothetical example in this way.[1]

An alternative approach to the payoff matrix is to consider the opportunity loss,[2] L_{ij}, associated with a given outcome rather than the V_{ij}. The *opportunity loss*, or *regret* matrix is calculated as follows:

1. For each S_j determine the maximum V_{ij} and denote this value by V_j^*.
2. The opportunity loss, L_{ij}, is then computed by

$$L_{ij} = V_j^* - V_{ij}$$

[1] Such complication might be very worthwhile in an actual decision situation. An alternative to increased calculation in such cases is to rely on a prior distribution which can be approximated by a mathematically based probability density function. We explore this alternative in Chapters 5 and 6.

[2] The reader will recall from Chapter 3 that the opportunity loss for a given outcome is defined to be the difference between the payoff of that outcome and the payoff from the best decision which could have been made for the state of nature which actually occurred.

Table 4-2 presents the opportunity-loss matrix for the franchise-expansion problem.

TABLE 4-2

Franchise-Expansion Problem—Opportunity-Loss Matrix (dollars)

Course of Action, A_i	State of Nature, S_j		
	S_1	S_2	S_3
A_1: Retain present facilities	0	900	1,800
A_2: Expand present facilities	1,500	0	800
A_3: Relocate and expand facilities.	3,000	1,500	0

In interpreting this table, the reader should keep in mind the relationship between opportunity loss and payoff, L_{ij} and V_{ij}. If, for example, S_3 were the true state of nature, the decision maker should have chosen A_3, since V_{33} = \$3,000 is greater than any other payoff associated with S_3. If our manager had chosen A_2 rather than A_3, his payoff would have been \$2,200 instead of \$3,000. Consequently, he would "regret" missing the opportunity to earn an additional \$3,000 - \$2,200 = \$800. Thus the opportunity loss, L_{23}, resulting from the intersection of A_2 and S_3 is \$800.

Given the data summarized in Tables 4-1 and 4-2, we can analyze the franchise-expansion problem. The decision maker can do one of three things. First, he can make a terminal decision—that is, choose a course of action, using the information now available (prior analysis). Second, he can evaluate the alternative of collecting additional information and then making a terminal decision (preposterior analysis). Third, he can decide now to collect additional information and then make a terminal decision (posterior analysis). Prior and preposterior analyses are discussed in this chapter; posterior analysis is considered in the next chapter.

PRIOR ANALYSIS

This section is concerned with *prior analysis*, making the terminal decision using available information, as opposed to assembling information specifically for this decision. Since the decision maker does not know which S_j will occur, he is making his decision under uncertainty. Prior analysis under uncertainty employs EMV and EOL as criteria of choice.

Expected Monetary Value

Let us suppose that, on the basis of his experience, judgment, and general knowledge of the situation, the decision maker estimates the probabilities to be .5, .3, and .2, for the occurrence of S_1, S_2, and S_3, respectively. The decision maker feels that there are five chances in ten that potential demand is only \$5,000, and only two chances in ten that this potential is as high as \$30,000.

These probabilities are used to calculate the EMV of each A_i. The results are summarized in Table 4-3. An efficient way

TABLE 4-3

Franchise-Expansion Problem Prior Analysis—Expected Monetary Value (dollars)

Course of Action, A_i	State of Nature, S_j			Probability of S_j		Expected Monetary Value	
	S_1	S_2	S_3				
		(V_{ij})		$\cdot\ P(S_j)$	=	EMV $[A_i]$	
A_1	1,000	1,100	1,200	.5		\$1,070	(max)
A_2	-500	2,000	2,200	.3		790	
A_3	-2,000	500	3,000	.2		-250	

of making these calculations is to multiply V, the payoff matrix, by P, the probability matrix, to get the expected monetary value matrix;

$$(V_{ij}) \cdot P(S_j) = \text{EMV}\ [A_i]$$

The decision rule is to select the A_i which results in the largest EMV $[A_i]$. For the present situation, since max (EMV $[A_i]$) = EMV $[A_1]$ = \$1,070, the decision maker should choose A_1.

The right-hand column of Table 4-3 indicates that the EMV of the courses of action are \$1,070, \$790 and -\$250, respectively. If the terminal decision were to be made on the basis of this prior information, the manager should reason in the following way. Since \$1,070 is the maximum of these expected values, A_1 is the optimum decision. That is, the decision maker should retain present facilities, and expect his total profit to average \$1,070 per month.

This analysis can also be explained by means of a decision tree. As discussed in Chapter 2, there are three steps in constructing a decision tree: (1) sketch all branches, (2) enter the relevant information on the proper branches, and (3) compute the expected values at each node. For the present problem, the tree has three main branches—one for each course of action. Each decision branch has three branches corresponding to the three states of nature. The relevant data are the probabilities of each state of nature, $P(S_j)$, and the payoffs associated with each outcome, V_{ij}. EMV's can then be computed from right to left and posted to appropriate nodes on the tree. The optimum

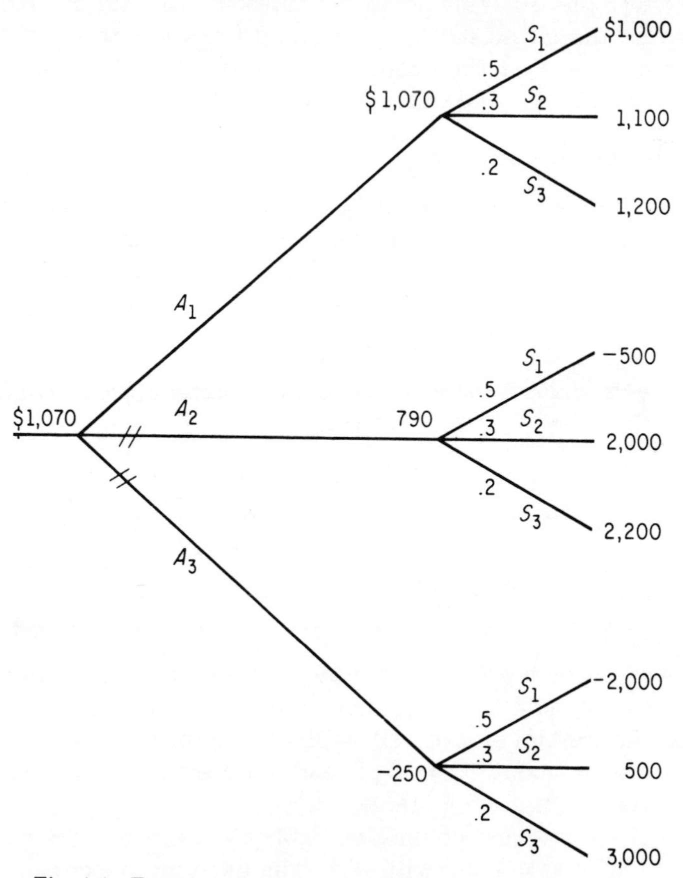

Fig. 4-1. Franchise-expansion problem—prior analysis by decision tree.

decision path is then marked by "blocking" nonoptimum branches with a double slash. The optimum path can then be read from left to right. Figure 4-1 is the decision tree for the prior analysis of our decision problem, using EMV as the decision criterion.

A value appears at each node of the tree. Where the tree branches from A_i to S_j, the nodal value is determined by computing the EMV $[A_i]$. Where the tree branches to decision alternatives, the nodal value is determined by taking the maximum EMV $[A_i]$ leading to that node. Thus, for A_1 the EMV is $1,070. Since the EMV $[A_1]$ is larger than any other EMV $[A_i]$, the other branches are blocked off, and $1,070 is posted at the leftmost node. Reading from left to right, the reader can see that the decision maker should follow the top branch, choose A_1, and expect an average profit of $1,070.

Expected Opportunity Loss

An alternative method of prior analysis is to calculate the EOL associated with each action. The opportunity-loss concept, as applied to the data of our franchise-expansion problem, is summarized in Table 4-4. Calculation of the EOL for each

TABLE 4-4

Franchise-Expansion Problem Prior Analysis—Expected Opportunity Loss (dollars)

Course of Action, A_i	State of Nature, S_j			Probability of S_j		Expected Opportunity Loss	
	S_1	S_2	S_3				
		(L_{ij})		$\cdot$ (P_j)	=	EOL	
A_1	0	900	1,800	.5		$630	(min)
A_2	1,500	0	800	.3	=	910	
A_3	3,000	1,500	0	.2		1,950	

course of action is accomplished by multiplying the opportunity loss for each outcome, V_{ij}, times the probability of that outcome and summing over all S_j's. Given prior probabilities $P(S_j)$ of .5, .3, and .2 for $j = 1$, 2, and 3 respectively, the EOL of each action A_i, is given in Table 4-4.

As the right-hand column of Table 4-4 indicates, the minimum EOL is associated with A_1. The decision maker can expect that, on the average, his opportunity loss will be lower if he

'etains present facility than if he expands or expands and re-
ocates. If the manager were going to make his decision on the
basis of the prior information only, he would choose A_1. The
EOL criterion and the EMV criterion indicate that the same
decision is optimal. As noted in Chapter 3, this result is not
coincidental; it reflects the functional relationship between the
payoff and loss matrices. Decision-tree analysis could also employ opportunity losses rather than conditional payoffs.

The reader will note that even the best decision for this
situation has an expected opportunity loss with a present value
of \$630. This amount is a function of the decision maker's
uncertainty concerning the state of nature which will occur.
The minimum EOL is the inherent cost of uncertainty itself.
It cannot be reduced by better decision making. The only way
the cost of uncertainty can be reduced is to obtain additional
information about the occurrence of states of nature in the
decision situation. The next section discusses the value of such
information.

PREPOSTERIOR ANALYSIS

The assembly of additional information is not without cost.
The decision maker should not waste time and money simply
to collect data. If and only if the value of the additional information exceeds the total of the assembly costs and the expected
value of the best prior decision should that information be assembled. The process of determining whether or not data for a
posterior decision (decision made after additional information)
should be assembled is termed *preposterior analysis.*

Expected Value of Perfect Information

The upper bound or limit on the value of additional information can easily be seen. If there were no uncertainty as to
which state of nature would occur, the decision maker would always choose the course of action which maximized the payoff
for the state of nature known to occur. The EOL would be
zero. Consequently, the upper bound on the value of additional
information is the cost of uncertainty: the minimum EOL. In
the present problem, this amount is \$630. This upper bound is
termed the *Expected Value of Perfect Information*, or EVPI.

The EVPI can also be calculated using EMV. To do this, the concept of "Expected Value Under Certainty" or EVUC must be explained. This amount is defined as the average profit which the decision maker could expect to earn if, in an infinitely long series of repetitions of the decision situation being studied, the occurrence of the three states of nature varied according to the decision maker's prior probability distribution, and the decision maker knew in advance which state would occur on a given repetition. Note that this concept is defined by the prior probability distribution; it is not the payoff which would be obtained if the decision maker had "perfect knowledge" of the future situation. As Schlaifer remarks, "this latter figure is known to God alone and is irrelevant to the problem of decision making under uncertainty."[3]

For our franchise-expansion problem, the EVUC is calculated as follows:

1. Determine the maximum payoff, V_j^*, for each state of nature.
2. Multiply each maximum payoff for a given state of nature by the probability of occurrence of that state.
3. Sum over all states.

Symbolically,

$$\text{EVUC} = \sum_{j=1}^{n} V_j^* \cdot P(S_j)$$

For our franchise-expansion problem,

$$\text{EVUC} = [\$1{,}000, \$2{,}000, \$3{,}000] \cdot \begin{bmatrix} .5 \\ .3 \\ .2 \end{bmatrix} = \$1{,}700$$

After the Expected Value Under Certainty has been calculated, the maximum EMV must be subtracted, since this amount will be earned, on the average, with no additional information. The difference is the cost of uncertainty, the minimum EOL, or the EVPI. This relationship can be summarized as follows:

$$\text{EVPI} = \text{EVUC} - \max \text{EMV} = \min \text{EOL}$$

[3] Robert Schlaifer, *Probability and Statistics for Business Decisions* (New York: McGraw-Hill Book Company, 1959), p. 122.

In terms of the present problem,

$$\text{EVPI} = \$1{,}700 - \$1{,}070 = \$630$$

The EVPI thus sets an upper limit on the amount which the decision maker would be willing to pay for additional information. Unfortunately, information is rarely perfect. Errors occur in collecting and tabulating information. If sampling is employed, sampling error will also occur. Competitors may try to distort the information. Time pressures may preclude a complete examination. The list of error possibilities is virtually endless. Consequently, the actual expected value of additional information is almost certain to be less than the EVPI.

Revision of Prior Probabilities

To determine how much he should pay for "imperfect information," the decision maker needs to know the degree of imperfection. Suppose he is approached by a marketing research firm which offers to study the market and tell him which state of nature will occur. The firm will charge $100 for this study, and estimates that it will be 70 percent reliable. That is, if the survey shows that a given state of nature will occur, the probability of that state actually occurring is .70. The probability that the survey is incorrect, i.e., that it indicates a state other than the one which actually occurs is 1.00 - .70 = .30. The survey is incorrect if it shows either of the states of nature which do not occur. If the manager assumes the error is equally likely for either incorrect state, the probability of the survey showing a particular state, and having some other state occur is .30/2 = .15.

These conditional probabilities can be summarized as follows:

$$P(R_k \mid S_j) \begin{cases} .70 & \text{for } k = j \\ .15 & \text{for } k \neq j \end{cases}$$

where R_k = kth survey result (k = 1, 2, 3)

S_j = jth state of nature which actually occurs (j = 1, 2, 3)

This statement is read, "the probability that the survey shows that state of nature k will occur, given that state of nature j actually occurs, is .70 when the survey is correct, and .15 when it is incorrect." For example, $P(R_1 | S_1) = .70$ means the probability that the survey shows state of nature S_1 will occur, given that state of nature S_1 does occur, is .70. $P(R_2 | S_1) = .15$ means the probability that the survey shows state of nature S_2 will occur given that state of nature S_1 does occurs is .15. These probabilities can also be presented in tabular form as shown in Table 4-5.

TABLE 4-5

Franchise-Expansion Problem—Conditional Probabilities of Survey Results

Survey Result, R_k	State of Nature, S_j		
	S_1	S_2	S_3
R_1	.70	.15	.15
R_2	.15	.70	.15
R_3	.15	.15	.70

To determine whether the proposed survey is worth its cost of $100, the decision maker must use these conditional probabilities to revise his prior probabilities, and recalculate the EMV of each action using the revised probabilities. If the EMV (computed using the revised probabilities) exceeds the cost of the survey plus the EMV (computed using the prior probabilities) without the survey, the manager should take the survey. If not, he should make his terminal decision on the basis of the prior analysis.

Revision of the prior probabilities is accomplished by using Bayes' theorem. Bayes' theorem may be restated for the present problem as

$$P(S_j | R_k) = \frac{P(R_k | S_j) \cdot P(S_j)}{P(R_k)}$$

This statement is read, "the probability that state of nature j will occur, given that the survey shows that state k will occur, is equal to the conditional probability of survey result k given state j times the marginal probability that state j occurs divided by the marginal probability that the survey will show result k." This calculation must be performed for each of the three possible survey results given each of the three possible states of nature. The results of these calculations are summarized in Table 4-6.

TABLE 4-6

Franchise-Expansion Problem—Revision of Prior Probabilities

urvey esult, R_k	State of Nature, S_j			Marginal Probability of Survey Result $P(R_k)$	Revised Probabilities		
	S_1	S_2	S_3		$P(S_1 \mid R_k)$	$P(S_2 \mid R_k)$	$P(S_3 \mid R_k)$
R_1	.350	.045	.030	.425	.823	.106	.071
R_2	.075	.210	.030	.315	.238	.667	.095
R_3	.075	.045	.140	.260	.289	.173	.538
$P(S_j)$	.500	.300	.200	1.000			

The entries in the first three columns of Table 4-6 are the joint probabilities of the occurrence of a particular survey result and a specific state of nature. They are calculated by multiplying the conditional probability of the occurrence of the particular survey result given the (subsequent) occurrence of the specific state times the marginal (prior) probability of the occurrence of that state. For example, the first entry, .350, is obtained as follows:

$$P(R_1 \cap S_1) = P(R_1 \mid S_1) \cdot P(S_1) = (.70)\,(.50) = .350$$

Notice that, for a given state, the joint probabilities sum vertically to the marginal probability of the occurrence of that state. Each entry in the fourth column is the horizontal sum of the first three rows. This sum is the marginal or unconditional probability of the occurrence of each of the possible survey results. For example, the unconditional probability that the survey will show state of nature two is .315. Notice that the sum of this column is 1.000, which means that all possibilities are included, and that this total checks with the sum of the marginal probabilities of each of the three states of nature.

Each entry in the last three columns is the revised probability of the occurrence of a specific state, given the additional information that the survey has shown that a particular state will occur. For example, using Bayes' theorem, the first entry in column 5, .823, is given by

$$P(S_1 \mid R_1) = \frac{P(R_1 \mid S_1) \cdot P(S_1)}{P(R_1)} = \frac{.350}{.425} = .823$$

This figure is the probability that state of nature one will occur given that the survey showed that it would. This probability is a revision of the prior probability of .500. The revision is upward, since the additional information supports the prior information. The second entry in this column, .238, is the downward revision of the .500 prior probability of occurrence of state 1, given that the survey showed that state 2 would occur. The other entries are similarly calculated and interpreted.

Calculation of Expected Net Gain of Sampling

These revised probabilities are used to recalculate the EMV of each of the three courses of action. The results are presented below. To obtain the revised EMV matrix, denoted Z_{ik}, we multiply the payoff matrix, V_{ij}, by the matrix of conditional probabilities, C_{jk}.[4]

$$\underset{V_{ij}}{\begin{bmatrix} 1{,}000 & 1{,}100 & 1{,}200 \\ -500 & 2{,}000 & 2{,}200 \\ -2{,}000 & 500 & 3{,}000 \end{bmatrix}} \cdot \underset{C_{jk}}{\begin{bmatrix} .823 & .238 & .289 \\ .106 & .667 & .173 \\ .071 & .095 & .538 \end{bmatrix}}$$

$$= \underset{Z_{ik}}{\begin{bmatrix} 1{,}024.80 & 1{,}085.70 & 1{,}124.90 \\ -43.30 & 1{,}424.00 & 1{,}385.10 \\ -1{,}380.00 & 142.50 & 1{,}122.50 \end{bmatrix}}$$

The resulting matrix Z_{ik} is the matrix of revised EMV's associated with each of the three survey results. For example, Z_{32} = \$142.50 is the expected payoff resulting from the third action, A_3, and the second survey result, R_2. The EMV of the survey is then determined by taking the expectation of the three survey results. This is done by selecting the best action for each survey result and multiplying the payoff for that action by the marginal probability of the occurrence of that survey result. Let $Z_k^* = \max(Z_{ik})$ denote the best action for the kth survey result. This selection produces a matrix $Z_k^* =$ [1,024.80, 1,424.00, 1,385.10] of maximum payoffs of action given the kth survey result. The matrix of maximum payoffs,

[4]Note that since multiplication is performed across the j states of nature, the conditional probability matrix, C_{jk}, is the *transpose* of the matrix of revised probabilities in Table 4-6.

Z_k^*, times the matrix of marginal probabilities of survey results, $P(R_k)$, gives the EMV $[S]$. The calculations are

$$Z_k^* \cdot P(R_k) = \text{EMV}[S]$$

$$[1{,}024.80, 1{,}424.00, 1{,}385.10] \cdot \begin{bmatrix} .425 \\ .315 \\ .260 \end{bmatrix} = \$1{,}244.23$$

The final step is to subtract the cost of the survey from EMV$[S]$, to obtain the Expected Net Gain of the Survey, ENGS = \$1,244.23 - \$100.00 = \$1,144.23. Since ENGS is greater than max EMV $[A_i]$, i.e., \$1,144.33 > \$1,070.00, the survey should be commissioned, and the survey results should be integrated with the results of the prior analysis to determine the appropriate decision.

Preposterior Analysis by Decision Tree

An alternative method of presenting results of the preposterior analysis is to construct a tree diagram. The completed tree is shown in Fig. 4-2. The first step in the construction is to sketch the branches. There are three possible survey results. For each possible result, the decision maker can choose one of three courses of action. And for each action there are three possible states of nature. Consequently, there are $3 \cdot 3 \cdot 3 = 27$ possible outcomes of this situation.

The second step is to attach to each outcome the proper conditional payoff V_{ij}, and the proper revised probability $P(S_j | R_k)$. For example, the first survey result, the third course of action and the second state of nature produce V_{32} = \$500 and $P(S_2 | R_1) = .106$. The reader should locate the branch corresponding to these figures, and trace through several other examples to be sure that this step is understood.

The third step in constructing this tree is to perform the necessary calculations and enter them in the proper positions on the tree. Continuing the example from the paragraph above, (\$500)(.106) = \$53. Performing similar calculations for nodes (A_3, S_1) and (A_3, S_3) yields -\$1,646 and \$213 respectively. Summing these three values gives Z_{31} = -\$1,380, the expected value of A_3 for R_1. This procedure is followed for the other A_i branching from R_1. These values are \$1,024.80 and -\$43.30 for A_1 and A_2 respectively.

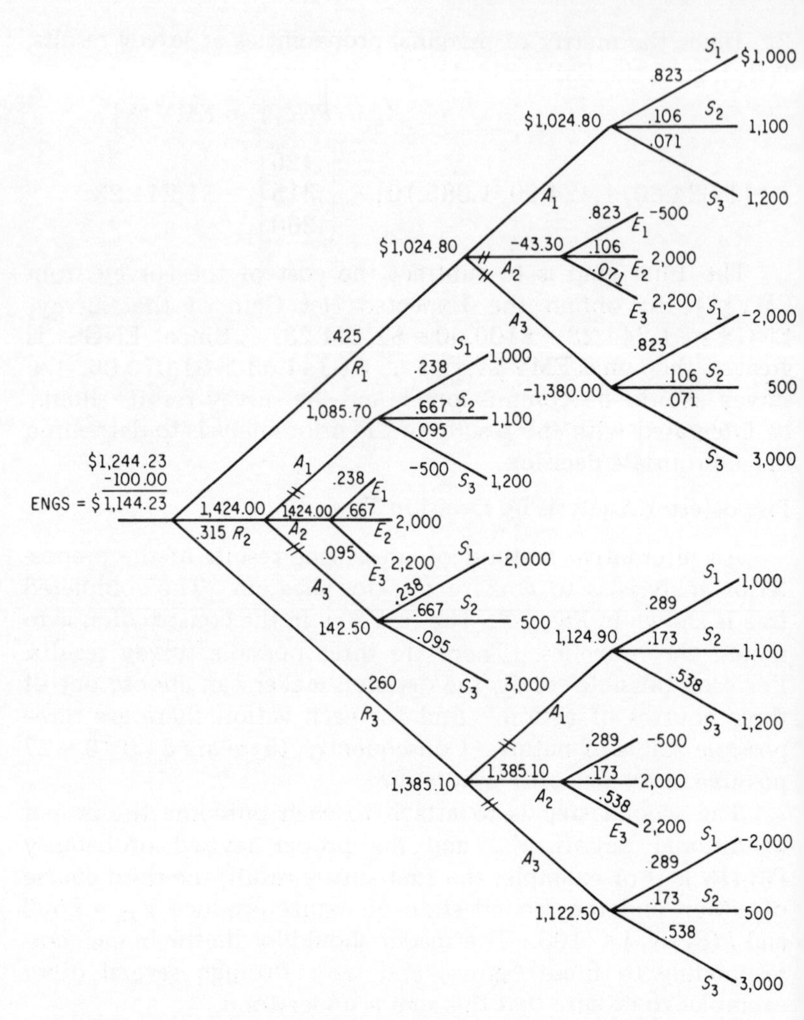

Fig. 4-2. Franchise-expansion problem—preposterior analysis by decision tree.

To obtain the value for node R_1, the value Z_1^* is determined by selecting the largest Z_{i1}. For R_1, this value is Z_{11} = $1,024.80. After nodal values for each R_k have been determined in this manner, the expected value of the survey is calculated by multiplying each Z_k^* by $P(R_k)$, the marginal probability of the kth survey result. We have

$$[1{,}024.80, 1{,}424.00, 1{,}385.10] \cdot \begin{bmatrix} .425 \\ .315 \\ .260 \end{bmatrix} = \$1{,}244.23$$

Subtracting the cost of the survey, $100, from this value leaves $1,144.23, the net expected monetary value to be derived from the additional sample information, i.e., ENGS. Since this expected value is greater than the maximum expected value without sampling, the decision maker should take the survey, given that he uses EMV as his decision criterion. Similar calculations using EOL as the criterion of choice would result in the same decision.

RECAPITULATION OF THE BAYESIAN DECISION PROCESS

It is appropriate at this point to recapitulate our discussion of the fundamentals of Bayesian analysis. This recapitulation will take two forms: (1) the decision model presented in Chapter 1, and (2) a decision-tree linkage of the prior and preposterior analyses of the franchise-expansion problem.

The Bayesian Decision Model

The decision model presented in Chapter 1 is reproduced in Fig. 4-3, with the addition of the concepts and procedures developed in this chapter which are necessary to operationalize the model. The problem is defined in terms of monetary payoffs, the V_{ij}. For the franchise-expansion example, the payoff matrix, V_{ij}, is given in Table 4-1. The prior analysis is performed by calculating the EMV of each course of action, EMV $[A_i]$. This result was

$$\text{EMV}[A_i] = \begin{bmatrix} \$1{,}070 \\ 790 \\ -250 \end{bmatrix}$$

for $i = 1, 2,$ and 3, respectively. If the decision were to be made on the basis of the prior analysis, the rational decision maker would choose A_1, since the EMV $[A_1]$ was the maximum EMV $[A_i]$.

Normally, however, the terminal decision should not be made without considering the potential value of additional in-

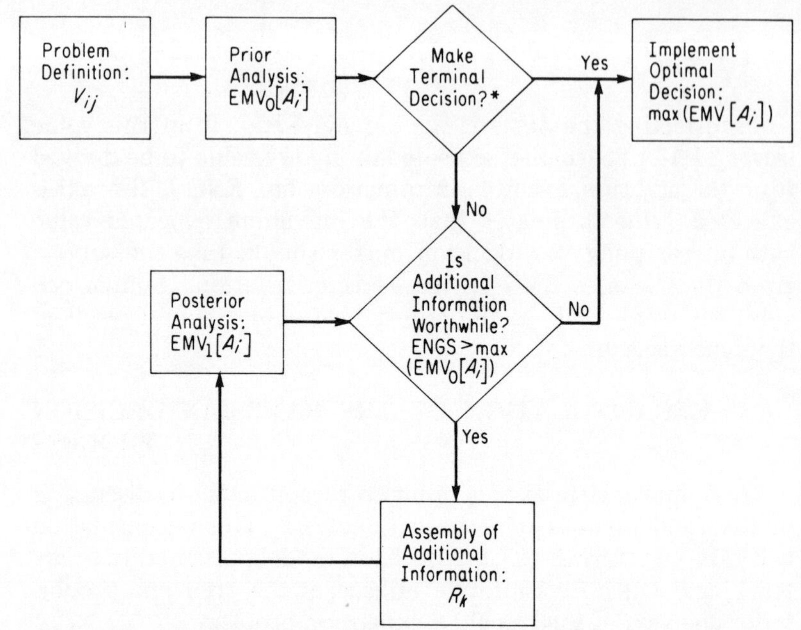

* Terminal decision made with no consideration of the value of additional information only if a dominant alternative exists.

Fig. 4-3. Review of Bayesian decision process.

formation. Only when a dominant course of action exists can the decision maker be sure that additional information will be of no value.[5]

When no dominant course of action exists, the value of additional information should be assembled when its expected net gain, ENGS, is greater than the maximum EMV obtained in the prior analysis, max EMV_0 $[A_i]$. (The reader will note that a zero has been subscripted to the EMV; this denotes prior analysis). Of course, if the expected net gain of the survey information is less than maximum EMV of the prior analysis, then the course of action having the maximum EMV should be selected with no further analysis.

If additional information is worthwhile, then it will be as-

[5] The reader is reminded that a dominant course of action is defined to be one which has a payoff as high or higher than any other alternative for all states of nature; see p. 44.

sembled. We have, in this book, ignored many interesting and significant problems associated with information assembly in order to concentrate on the decision process.[6] We have assumed that, for the franchise-expansion problem, a "survey" could be taken, and that possible survey results are denoted by R_k, where k is a subscript indicating state of nature. Additional information was found to be worthwhile for this problem.

Once this information has been assembled, it is integrated with the prior information and evaluated. This procedure is termed *posterior analysis.* The result of the posterior analysis is a revised set of EMV's for the courses of action: $\text{EMV}_1\,[A_i]$. The maximum EMV from this calculation indicates the best action, unless a new set of additional data would be worthwhile. If more data are to be collected, the EMV calculations become prior information, i.e., the $\text{EMV}_1\,[A_i]$ becomes $\text{EMV}_0\,[A_i]$ for the second round. The new $\max(\text{EMV}_0\,[A_i])$ is compared to a new ENGS. If this second round of additional information is found to be worthwhile, it is assembled. Then the new posterior expected monetary values can be calculated.

The process of comparing the EMV's to the ENGS and assembling and reanalyzing the data continues until additional information is not worthwhile. Then the course of action having the highest EMV is implemented. The $\max(\text{EMV}[A_i])$ in the implementation box in the model is not subscripted, since this expected value can result from either prior or posterior analysis.

Decision-Tree Linkage of Prior and Preposterior Analyses

Decision trees were utilized in both the prior and preposterior analyses of the franchise-expansion problem. In this section, the relationship between these analyses is shown by

[6] Two textbooks which treat these problems are H. W. Boyd, Jr. and Ralph Westfall, *Marketing Research; Text and Cases*, rev. ed. (Homewood, Ill.: Richard D. Irwin, Inc., 1964), and P. E. Green and D. S. Tull, *Research for Marketing Decisions* (Englewood Cliffs, N.J. Prentice-Hall, Inc., 1966). A forthcoming book, Keith K. Cox and Ben M. Enis, *The Marketing Research Process: Purchasing Relevent Information for Decision Making* (Pacific Palisades, Calif. Goodyear Publishing Company) approaches these problems within the framework of the Bayesian decision process outlined in this book.

means of a third decision tree. Figure 4-4 presents this relationship.

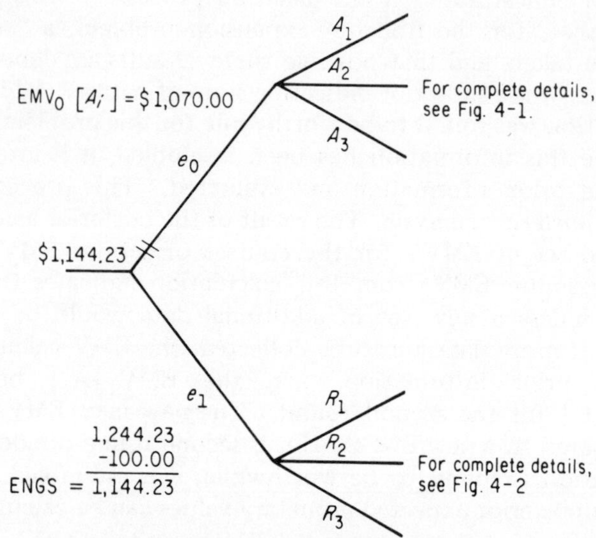

Key: e_0 is the event that no survey is taken and e_1 is the event that the 70 percent reliable survey is taken.

Fig. 4-4. Partial decision tree of Bayesian analysis —franchise-expansion problem.

The prior analysis, detailed in Fig. 4-1, indicated that A_1 was the best action and that the EMV of this decision was $1,070. The preposterior analysis (Fig. 4-2) showed that the ENGS was $1,144.23. The decision maker can now decide whether or not the survey should be commissioned. Since the ENGS is greater than the max(EMV$_0$ [A_i]), the rational decision maker would choose to commission the survey. Accordingly, the branch labeled e_0 (do not take survey) is blocked; the decision path is e_1, take the survey. For this reason, the terminal decision cannot be made until the posterior analysis is completed. This topic is the subject of the following chapter.

SUMMARY

This chapter presents the basic ideas of Bayesian analysis: payoff matrix construction, prior analysis of courses of action, and preposterior analysis. The payoff matrix is constructed from the basic decision model $V_{ij} = f(A_i, S_j)$ after the specific form of this function has been determined for the situation at hand. This model summarizes the courses of action available to the decision maker.

The decision maker analyzes these data, using either EMV or EOL as his decision criterion. On the basis of this analysis, he can either make the terminal decision, or evaluate the possibility of obtaining additional information. The former course is termed *prior analysis*, since the decision is made before any additional information is assembled. After additional data are collected, the analysis is termed *posterior analysis.*

The purpose of assembling additional information is to reduce the uncertainty as to which state of nature will occur. This reduction should, of course, be of greater value than the cost of assembling the information. This condition should be checked before the data are assembled. The checking procedure is termed *preposterior analysis*, because it is a calculation preliminary to the posterior analysis. Preposterior analysis employs Bayes' theorem to revise the decision maker's prior probability distribution, given certain facts about the additional data. The revised probability distribution is then used to recalculate the EMV (or EOL) of each course of action.

The following chapters extend this basic Bayesian analysis. In Chapter 5, posterior analysis is covered, and this analysis is compared to the "classical" statistician's approach to decision making. This discussion is based on the binomial distribution. Chapter 6 is devoted to the role of the normal distribution as a prior distribution, and as a posterior distribution, in Bayesian analysis. Chapter 7 summarizes the book, from a managerial perspective, including a discussion of limitations and examples of applications of Bayesian analysis in marketing, and discusses the future of Bayesian analysis for marketing decision making.

REFERENCES

Alderson, Wroe, and Paul E. Green. *Planning and Problem-Solving in Marketing.* Homewood, Ill.: Richard D. Irwin, Inc., 1964, Part II.

Part II is Green's masterful exposition of Bayesian analysis in marketing; nonmathematical, but "meaty."

Raiffa, Howard. *Decision Analysis.* Reading, Mass.: Addison-Wesley, Inc., 1968.

Very readable introduction to Bayesian analysis; requires no background in mathematics.

Schlaifer, Robert. *Probability and Statistics for Business Decisions.* New York: McGraw-Hill Book Company, 1959.

The standard reference for the application of Bayesian analysis to business problems; should be studied by every serious student of decision theory and marketing management.

PROBLEMS

4-1. The following payoff matrix is given (millions of dollars):

Courses of Action	States of Nature	
	S_1	S_2
A_1	8	1
A_2	4	3
A_3	6	1

(a) Given $P(S_1) = .3$ calculate:
  1. Expected monetary value
  2. Expected opportunity loss
  3. Expected profit under certainty
  4. Expected value of perfect information

(b) Calculate the probability of occurrence of S_2 which results in $\text{EMV}(A_1) = \text{EMV}(A_2)$.

(c) Suppose that a 70 percent reliable survey could be taken to determine which state of nature would occur.
  1. Use this information to revise the probabilities of occurrence of S_1 and S_2.
  2. Construct a decision tree showing the expected monetary value of using and of not using the survey information. Is this information worth the asking price of $.4 million?

4-2. Mr. John Shaw, Marketing Vice-President of CBA manufacturing company, is considering the expansion of his product into a new territory. Three distribution alternatives are available: a regular wholesaler, a manufacturer's agent, or setting up his own sales office. To use the wholesaler would cost 35 percent of sales revenue. The agent would cost $12,000 per year plus 15 percent of sales. Setting up a company sales office would require a flat $25,000 per year.

Of course, the "best" of these alternatives depends upon the amount of the product which can be sold each year. Market potential for the product in this territory is $400,000 annually. Mr. Shaw believes that sales of 10 percent of this market would be "low," a 20 percent share would be "medium," and a 30 percent share would be "high." He feels that the odds on achieving these shares are 3 in 10, 5 in 10, and 2 in 10 respectively. To keep it simple, assume that all other costs (manufacturing, general overhead, storage, etc.) amount to 60 percent of sales.

(a) Construct a payoff matrix.

(b) Using EMV as your decision criterion, perform a prior analysis. Calculate EVPI. What decision should be made?

(c) Suppose that an 80 percent reliable survey (with error possibilities equally divided) were available for $150. Revise Mr. Shaw's prior probabilities, and perform a posterior analysis. What decision should be made?

(d) Construct a tree diagram of the decision situation analyzed in this problem.

CHAPTER 5

Bayesian Analysis: The Binomial Distribution

This chapter extends the previous discussion of Bayesian analysis. First, the prior and preposterior analyses of the franchise-expansion problem introduced in Chapter 4 will be modified to more clearly illustrate the role of the binomial distribution in Bayesian analysis. Next, the nature of the binomial probability mass function is discussed. Then posterior analysis, using binomial probabilities, is performed. Finally, Bayesian analysis is compared to the classical "hypothesis-testing approach" to decision making.

MODIFICATION OF THE FRANCHISE-EXPANSION ANALYSIS

Consider a modified version of the franchise-expansion problem with the decision maker, as before, confronted with the question: "Should present facilities be retained or expanded?" Since expansion involves some cost, the answer depends on the revenues generated by the alternative courses of action. If price remains constant, revenue is a function of the proportion of prospective customers in the market area who prefer to patronize the franchiser. The higher this proportion, the more profitable expansion becomes relative to retaining the present facilities.

We will again assume that the decision maker has revenue and cost data which permit him to specify the functional relationship between the conditional payoffs and the proportion of customers in the market area who prefer to patronize his establishment—that is, his market share.

The decision alternatives are retention of present facilities (A_1), and expansion of facilities (A_2)[1]. Consequently, the analysis can be simplified by considering the incremental impact of one of the courses of action on profits. This can be accomplished by setting the payoff from A_1 (retention of present facilities) to zero, and then comparing the incremental payoff from A_2 (expansion of facilities) to zero payoff.

Consequently, the incremental payoff relationship can be described as follows. If A_2 were chosen and no customers in the market area purchased the product, the franchise would

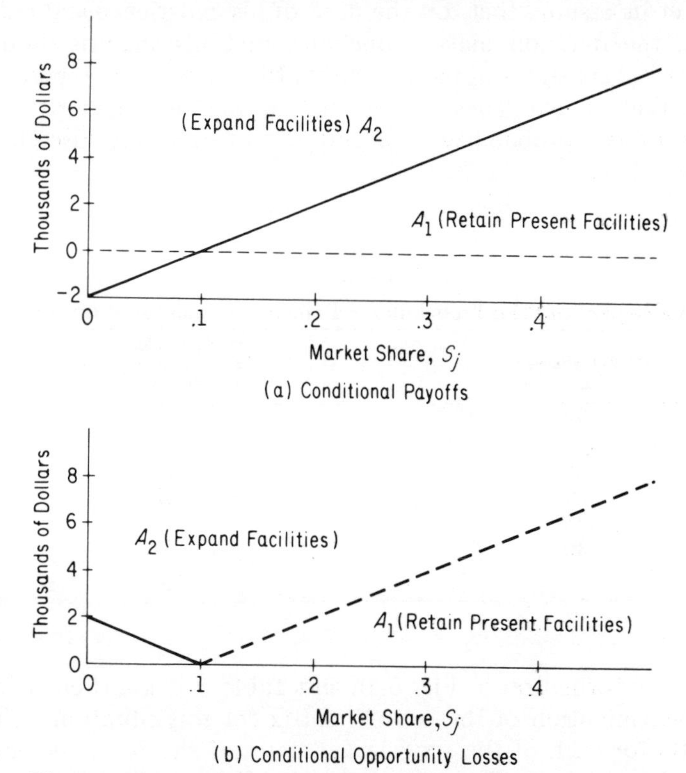

Fig. 5-1. Conditional payoffs and opportunity losses—franchise-expansion problem.

[1] The third alternative, relocation of the franchise, is not considered this analysis.

incur a $2,000 loss. (The reader will recall that all payoffs are stated in terms of present values of cash flow over the planning horizon.) A_1 is the preferred course of action until the proportion of customers who prefer the franchiser's product reaches 10 percent. If the true state of nature is that the proportion exceeds 10 percent, A_2 is the preferred course of action to follow. The conditional payoffs associated with A_2 increase as the proportion increases and reach a maximum of $6,000 at full capacity. Figure 5-1a illustrates these relationships. Figure 5-1b, which will be used later, is derived from Fig. 5-1a. It shows these incremental relationships in terms of conditional opportunity losses.

Let us assume that, on the basis of his experience and judgment, the decision maker concludes that his analysis should encompass six states of nature, within the range of 5–30 percent of market share. These states and the decision maker's estimates of the probability of occurrence of each state are given in Table 5-1.

TABLE 5-1

Assessment of Prior Probabilities—Franchise-Expansion Problem

Market Share, S_j	Decision Maker's Assessment of Prior Probabilities, $P(S_j)$
.05	.20
.10	.40
.15	.20
.20	.10
.25	.07
.30	.03
	1.00

The information in Fig. 5-1a and Table 5-1 is sufficient for the determination of the payoff matrix for this situation. The payoffs for each of the six relevant states of nature can be read from Fig. 5-1a. The resulting payoff matrix is given in Table 5-2.

TABLE 5-2

Payoff Matrix for Modified Franchise-Expansion Problem (thousands of dollars)

Course of Action, A_i	State of Nature, S_j					
	.05	.10	.15	.20	.25	.30
	s_1	s_2	s_3	s_4	s_5	s_6
A_1 (Retain present facilities) .	0	0	0	0	0	0
A_2 (Expand facilities).	−1	0	1	2	3	4

Using EMV as the decision criterion, we are ready to perform the prior analysis of this situation:

$$\text{EMV}_0\,[A_i] = \begin{bmatrix} 0 & 0 & 0 & 0 & 0 & 0 \\ -1 & 0 & 1 & 2 & 3 & 4 \end{bmatrix} \begin{bmatrix} .20 \\ .40 \\ .20 \\ .10 \\ .07 \\ .03 \end{bmatrix} = \begin{bmatrix} .00 \\ .53 \end{bmatrix}$$

If the terminal decision were to be made at this point, the decision maker should choose to expand facilities, since $\text{EMV}_0(A_2) = \$530$ is greater than $\text{EMV}_0\,[A_1] = \$0$.

Since there is some uncertainty as to whether this course of action is optimal for this situation, the decision maker might investigate the possibility of obtaining additional information. This preposterior analysis would proceed as follows. The Expected Value Under Certainty, obtained by summing the value of the best decision for each state of nature multiplied by the probability of occurrence of that state, is

$$\text{EVUC} = [0,\ 0,\ 1,\ 2,\ 3,\ 4] \begin{bmatrix} .20 \\ .40 \\ .20 \\ .10 \\ .07 \\ .03 \end{bmatrix} = \$730.00$$

and the Expected Value of Perfect Information is

$$\text{EVPI} = \text{EVUC} - \max \text{EMV}_0 = \$730.00 - \$530.00 = \$200.00$$

Since imperfect information would have less value, the decision maker would decide at this point to expand his facilities, unless he could obtain additional information rather inexpensively.

We have briefly covered the prior and preposterior analyses of this modification of the problem considered in more detail in Chapter 4. The reader should review these details if necessary. The situation was modified for this chapter to focus attention upon posterior analysis—that is, the integrated evaluation of both prior and additional information. Since we have assumed that the additional information is binomially distributed, the next section discusses the nature of the binomial distribution.

THE BINOMIAL PROBABILITY DISTRIBUTION

Many marketing problems involve situations in which there are only two possible outcomes. These outcomes occur at random and the probability of each occurrence is constant for each trial. Such processes are called *Bernoulli processes.* Outcomes of Bernoulli processes are often specified as successes or failures. The proportion of successes over a large number of trials is called a *parameter*, denoted by p, and is constant for a given Bernoulli process. For example, in surveys, consumers are asked questions which can be answered yes or no. For a given population and question, the likelihood of getting a yes response remains constant regardless of the number of consumers questioned.

The probability of occurrence of various Bernoulli outcomes can be calculated. Consider, for example, an experiment in which a coin is tossed. Let p = the probability of a success (heads) and $q = 1 - p$ the probability of a failure (tails). Then the probability that n trials yields r successes can be expressed as[2]

$$P\,(r \mid n, p) = {}_nC_r p^r q^{n-r} \tag{5-1}$$

Equation 5-1 is the generalized expression for a Bernoulli process; it is called the *binomial probability mass function* and is

[2] The symbol ${}_nC_r$ is read "combinations of n things taken r at a time," and is computed by $\dfrac{n!}{r!\,(n-r)!}$.

used in computing binomial probabilities. For example, the probability of obtaining exactly three heads in six tosses of a fair coin can be computed as follows:

$$P\left(r = 3 | n = 6, p = \frac{1}{2}\right) = \frac{6!}{3!\,(6-3)\,!}\left(\frac{1}{2}\right)^3\left(\frac{1}{2}\right)^3 = \frac{5}{16}$$

The generalized expression of the binomial probability mass function is used in computing binomial probabilities, but such computations are time consuming. Since the binomial is often used, tables of binomial values have been compiled. Appendix A is such a table. We will illustrate its use shortly.

Posterior Analysis Using Binomially Distributed Sampling Information

We are now ready to perform the posterior analysis. Let us again ignore the interesting and significant problems associated with determining sample size, cost, and reliability, and suppose that the decision maker has the results of a market survey indicating the market share he can expect to obtain. His task is to combine results of the survey and the prior information in order to make a terminal decision.

To illustrate how the market survey results are combined with the prior information, we continue to use the modified version of the franchise-expansion problem introduced above. Also, in order that the reader might become familiar with appropriate procedures for analyzing binomially distributed variables, we will assume that the market-survey results are based on a probability sample of 100 respondents taken from the population of potential customers in the market area. The respondents were asked whether they would or would not patronize the decision maker's establishment. Twelve of the one hundred respondents included in the survey indicated a preference for patronizing the franchiser's establishment. These results are assumed to be binomially distributed.

The first step in performing the posterior analysis is to revise the prior probability distribution, given the market survey results. Table 5-3 summarizes the calculations required to compute the posterior probabilities.

TABLE 5-3

Revision of Probabilities—Franchise-Expansion Problem

State of Nature, S_j	Probabilities			
	Prior	Conditional	Joint	Posterior
	$P(S_j)$	$P(r = 12 \mid S_j)$	$P(S_j) \cdot P(r = 12 \mid S_j)$	$P(S_j \mid r = 12)$
.05	.20	.0028	.00056	.0096
.10	.40	.0988	.03952	.6795
.15	.20	.0838	.01676	.2882
.20	.10	.0128	.00128	.0220
.25	.07	.0006	.00004	.0007
.30	.03	.0000	.00000	.0000
			.05816	1.0000

The entries in Table 5-3, beginning with the first column, can be interpreted as follows: (1) the states of nature considered to be relevant by the decision maker—i.e., the proportions in the population that might patronize the franchise; (2) the prior probabilities assigned to these states of nature by the decision maker; (3) the conditional probabilities that a sample of size 100 will yield twelve positive responses, given the states of nature listed in the first column; (4) the joint probabilities resulting from the intersection of each prior probability and the corresponding conditional probability; and (5) the posterior probabilities—that is, the probability that each state of nature is the true state of nature, given that a sample of size 100 yielded twelve positive responses.

The conditional probabilities shown in the third column of Table 5-3 can be computed using the binomial probability mass function. For example, the value in the first row, third column, can be computed as follows:

$$P(r = 12 \mid n = 100,\ S_j = .05)$$

$$= \frac{100\,!}{12\,!\,(100 - 12)\,!}\,(.05)^{12}\ (.95)^{88} = .0028$$

However, such work is considerably shortened by making use of Appendix A, the cumulative binomial probability distribution. For example, to locate the value of $P(r = 12 \mid n = 100,$ $S_j = 12)$, locate the page for $n = 100$, the column for $S_j = .05$, the row for $r = 12$, and read the value .0043 directly from the table. Since the table gives the cumulative binomial values—

that is, the probability of twelve or more responses, given $S_j = .05$ and $n = 100$, it is necessary to subtract the value for thirteen or more from .0043 to obtain the probability of exactly twelve responses. Symbolically, $P(r \geqslant 12) - P(r \geqslant 13) = P(r = 12)$. For our problem, .0043 - .0015 = .0028. Similarly, other values in the third column are found under the appropriate S_j's in the table for $n = 100$.

The fourth column contains the probability of the joint occurrence of the prior and conditional probabilities for each state of nature. These values are obtained by multiplication:

$$P(R_k \cap S_j) = P(R_k|S_j) \cdot P(S_j)$$

where $P(R_k) = P(r = 12|S_j)$ to simplify the notation and conform to the notation of Chapter 4. For example, the probability that the true market share proportion is 5 percent and the survey produced twelve positive responses is

$$P(r = 12 \cap S_j = .05) = (.20)\,(.0028) = .00056$$

The other joint probabilities are similarly calculated. Summing these values gives the marginal probability of the survey result, twelve positive responses, in this situation.

Column 5 contains the revised or posterior probability of the occurrence of each state. These values are obtained using Bayes' theorem:

$$P(S_j|R_k) = \frac{P(R_k|S_j) \cdot P(S_j)}{\sum_{j=1}^{n} P(R_k|S_j) \cdot P(S_j)}$$

For example, given the survey results, the revised probability that the true market share proportion is 5 percent is

$$P(S_j = .05\,|\,r = 12) = \frac{.00056}{.05816} = .0096$$

The other posterior probabilities are similarly calculated.

These posterior probabilities are used to recalculate the EMV of the two courses of action:

$$\text{EMV}_1[A_i] = \begin{bmatrix} 0 & 0 & 0 & 0 & 0 & 0 \\ -1 & 0 & 1 & 2 & 3 & 4 \end{bmatrix} \begin{bmatrix} .0096 \\ .6795 \\ .2882 \\ .0220 \\ .0007 \\ .0000 \end{bmatrix} = \begin{bmatrix} .0000 \\ .3274 \end{bmatrix}$$

Thus the posterior analysis shows that the expected monetary value of expansion is $324.70 when compared to retention of the present facility.

Consequently, the decision maker should expand facilities. Note that A_2 was the preferred course of action for both the prior and the posterior analysis. However, the $\text{EMV}_1[A_2]$ is less than $\text{EMV}_0[A_2]$, which means that the decision maker's optimistic prior estimates were reduced somewhat by the survey results.

The results of the posterior analysis could be treated as prior information for a possible second survey. To determine the value of this possibility, we recompute the EVPI:

$$\text{EVUC} = [0 \quad 0 \quad 1 \quad 2 \quad 3 \quad 4] \begin{bmatrix} .0096 \\ .6795 \\ .2882 \\ .0220 \\ .0007 \\ .0000 \end{bmatrix} = \$334.30$$

and

$$\text{EVPI} = \$334.30 - \$324.70 = \$9.60$$

Since it is unlikely that the manager could acquire worthwhile information for $9.60, he should make the terminal decision at this point. Using EMV as his decision criterion, he should expand the facility.

The reader should bear several points in mind as he reviews this analysis. First, the situation has been simplified to highlight the analytical process. Many of the complexities of real-world problems of this type have been ignored. Secondly, no attention was given to the sampling accuracy of the additional data. Questions about such factors as the size and representativeness of the sample, interviewing techniques, nonresponse rate, and the correlation between answers to a questionnaire and actual behavior would be relevant. Thirdly, it should be remembered that decisions should not be made entirely "by the numbers." Perhaps expected monetary value is not an appropriate decision criterion, or maybe significant factors were not assessed in dollars-and-cents terms. Bayesian analysis can reduce, but usually cannot eliminate, the uncertainty surrounding a particular decision situation.

COMPARISON OF BAYESIAN AND CLASSICAL ANALYSIS

A better understanding of the Bayesian approach to decision making can be gained by considering the course of action that would have been selected had we used a classical (hypothesis testing) approach to the franchise-expansion problem. A classical statistician would probably approach the problem by formulating and testing the hypothesis that the proportion in the population preferring to patronize the franchise is less than or equal to 10 percent: the indifference point between expansion and retention of present facilities. More formally, the classical statistician would test the following hypotheses:

$$H_0: \; S \leqslant .10$$
$$H_1: \; S > .10$$

H_0 is called the *null hypothesis.* If it is true, the decision maker should choose A_1, because retention of present facilities has the higher conditional payoff (as shown in Fig. 5-1). H_1 is called the alternative hypothesis; if it is true, the decision maker should choose A_2 because A_2 has a higher conditional payoff for any market share above 10 percent.

The reader may recall a diagram similar to the one in Table 5-4 from elementary statistics. This diagram illustrates the relationship between decision making and hypothesis testing.

If H_0 is true, and it is rejected, a Type I error is made. If H_0 is false, and it is accepted, a Type II error is made. The probability a statistician assigns to a Type I error is called the *alpha risk;* the probability he assigns to a Type II error is called a *beta risk.* The "risk" is that, because of sampling inaccuracy, the sample results differ from the true state of nature. In

TABLE 5-4

Consequences of Decision Making by Hypothesis Testing

	State of Nature	
Decision	H_0 is True	H_0 is False
Accept H_0	Correct Decision	Type II Error
Reject H_0	Type I Error	Correct Decision

practice, the statistician usually assigns a probability to the alpha risk (.05 or .01 are typically assigned) associated with $S \leqslant .10$ and investigates the beta risks associated with a Type II error for various values of S_j. Note that this practice tends to maintain the status quo. The new concept must prove its value.

Let us apply these concepts to the franchise-expansion problem. If an alpha risk of 5 chances in 100 is selected (by management) for $H_0 \leqslant .10$, the decision maker would not reject H_0 unless the number in a sample of size 100 who preferred to patronize the franchiser's facilities exceeded 15. This rule is determined by using Appendix A. The procedure is as follows. Find the section of the table for $n = 100$, follow the column for $S = .10$ until .0399 is reached, and read $r = 16$ from the left margin. The value .0399 is closest to .05. (A value of exactly .05 cannot be obtained since the binomial is a discrete distribution.)

Now let us examine the error consequences of the decision rule:

$$\text{Accept } H_0 \text{ if } r \leqslant 15$$
$$\text{Reject } H_0 \text{ if } r > 15$$

This rule says retain the present facilities unless at least sixteen positive responses are obtained from a sample of 100 consumers. If the proportion of customers preferring the franchiser's facilities is .10 or less, then, because of sampling error, about four times out of 100 the decision maker following the rule will reject H_0 when it is in fact true. A rule that so rarely rejects H_0 leads to a high probability of accepting H_0 when it is false.

Figure 5-2 illustrates the alpha and beta risks associated with the above decision rule. The reader may recall from elementary

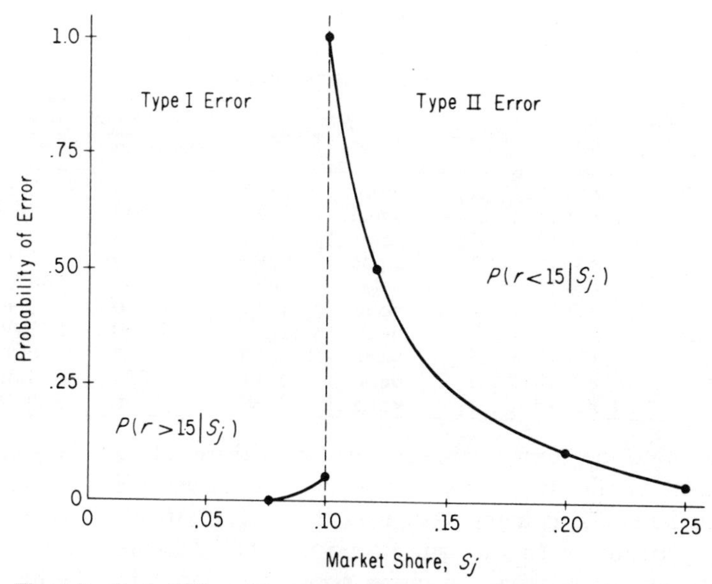

Fig. 5-2. Probability of error with hypothesis testing—franchise-expansion problem.

statistics that the portion of the curve corresponding to a Type I error is a portion of the "power curve," while the portion corresponding to a Type II error is a portion of the "operating characteristic" curve. Figure 5-2 shows that, given the hypothesis that $S \leqslant .10$ a low alpha risk is associated with a high beta risk. Also, as the alpha risk decreases, the beta risk increases.

Our next step is to investigate the economic consequences (opportunity losses) associated with making these erroneous decisions. Table 5-5 has been prepared for this purpose, as follows: (1) some values for the true population proportion were assumed (specifically, S_j = .09, .10, .15, .20, .25); (2) the probabilities of obtaining the specified sample result from a population with the assumed proportions were found in Appendix A; (3) the conditional opportunity losses were taken from Fig. 5-1b; and (4) the conditional expected opportunity losses were computed by multiplying each conditional loss by the probability of occurrence of that error.

The expected opportunity loss column of Table 5-5 is plotted in Fig. 5-3 to more graphically portray the economic consequences of the classical decision rule stated above. Note

TABLE 5-5

Expected Opportunity Losses of Classical Decision Rule—Franchise-Expansion Problem

Actual State of Nature (S_j)	Sample Outcome	Decision	Type of Error	Probability	Opportunity Loss: Conditional	Opportunity Loss: Expected
.09	$r \leqslant 15$	A_1	none	.9831	\$ 0	\$ 0.00
	$r > 15$	A_2	alpha	.0169	200	3.38
.10	$r \leqslant 15$	A_1	none	.9601	0	0.00
	$r > 15$	A_2	alpha	.0399	0	0.00
.15	$r \leqslant 15$	A_1	beta	.5683	1,000	568.30
	$r > 15$	A_2	none	.4317	0	0.00
.20	$r \leqslant 15$	A_1	beta	.1285	2,000	257.00
	$r > 15$	A_2	none	.8715	0	0.00
.25	$r \leqslant 15$	A_1	beta	.0111	3,000	33.30
	$r > 15$	A_2	none	.9889	0	0.00

that the opportunity losses associated with rejecting H_0 (retain present facilities) are low, while the cost of accepting H_0 (expand facilities) is high. Hence, if the decision maker attaches a low probability to the true proportion being greater than .10, he would be satisfied with the above decision rule. Perhaps, however, the decision maker would prefer a more balanced decision rule, i.e., one which accorded more equal treatment to the economic consequences of alpha and beta risks.

Since the sample result was that the proportion of re-

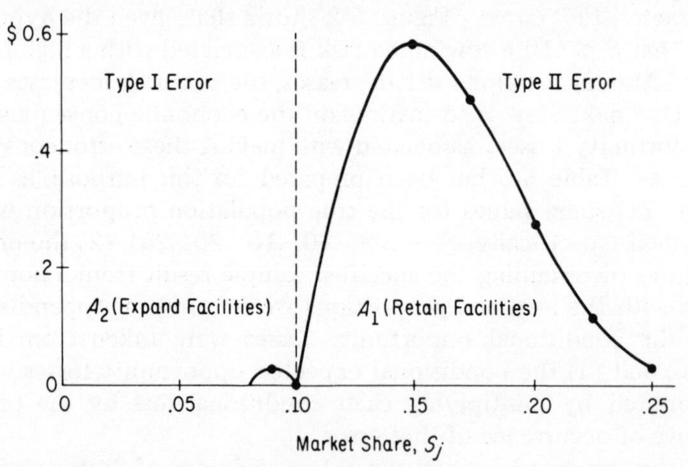

Fig. 5-3. Conditional expected opportunity losses associated with actions A_1 and A_2.

pondents who did favor the franchiser's expanded facilities was .2 percent, the classical statistician would conclude from his est that the true proportion was not significantly greater than .0 percent. Consequently, he would "accept the null hypothesis," i.e., would recommend retaining present facilities. He probably would not calculate the opportunity loss associated with this decision. It is the authors' position that, while the classical emphasis on preserving the status quo is appropriate n basic research (e.g., medicine) and in situations in which payoffs are difficult to calculate, explicit consideration should be given to the costs of incorrect economic decisions. Moreover, he classical approach ignores prior information which could be valuable in evaluating the decision situation.

SUMMARY

This chapter has discussed several extensions of Bayesian analysis. Posterior analysis, the integration of prior information and additional information, was illustrated using binomial probabilities. This analysis was compared to the classical approach o this problem. It was noted that the classical approach does not take explicit account of prior information, nor of opportunity losses. The next chapter discusses the role of the normal distribution in Bayesian analysis.

REFERENCES

(See also references for Chapter 4)

Chernoff, H., and L. E. Moses. *Elementary Decision Theory.* New York: John Wiley & Sons, Inc., 1959.

Well-written elementary treatment of decision theory, with particular attention to procedures for obtaining additional information about the state of nature.

Schlaifer, Robert. *Introduction to Statistics for Business Decisions.* New York: McGraw-Hill Book Company, 1961.

A shorter version of Schlaifer's 1959 classic, this book more explicitly contrasts Bayesian and classical approaches to decision making.

PROBLEMS

5-1. Consider the decision situation faced by the manager of a book club which sells paperback reprints of scientific books by direct mail only. He has learned that the publishing rights to a certain book are available. Production costs (payment for rights, setup of presses, etc.) would be \$3,000, plus a variable cost (paper, ink, labor, binding) of \$.50 per book. Club policy is to price its offerings at \$2 per book. Five thousand copies would be printed if the manager decided to print the book. His club has a mailing list of 55,000 customers. Of course, if he decides not to publish, he will experience no gain and no loss.

To get a rough idea of the economic consequences of this decision, the manager wishes to examine the results from three levels of response:

S_1 = 2 percent of the membership buys the book
S_2 = 5 percent buys
S_3 = 8 percent buys

His estimate of the probability of occurrence of each of these three states is .1, .3, and .6 for 2, 5, and 8 percent respectively.

(a) Show that the break-even market share for this situation is 5 percent.

(b) Set up the two-decision by three-states-of-nature payoff matrix for this situation.

(c) Graph the conditional opportunity loss function for each decision.

5-2. Suppose that the book club manager in question 1 would like to gather additional information. He would like to be 90 percent confident that he could at least break even on this book. He has decided that he will take a sample of 100 book club members. This sample will cost \$50 for printing of the mailing piece plus \$.24 per piece for processing and mailing.

(a) Calculate the expected value of perfect information, and show that the maximum expected value of this sample information would be \$256.

(b) Set up the classical decision rule for the null hypothesis implied in the manager's statement, and calculate the expected opportunity loss of this rule for both decisions given each of the three states of nature.

5-3. Suppose that the sample described in question 2 produced 4 orders. Given the assumption that the sampling result is binomially distributed, unbiased, and representative of the population of book-club members:

(a) What decision would the classical statistician advise the manager to make?

(b) Revise the prior probabilities given in question 5. Show that $P(S_j|r = 4) = .095, .565$, and $.340$ for $j = 1, 2, 3$ respectively.

(c) Calculate the posterior expected value of each decision. Using this decision criterion, which decision should the manager make?

5-4. Suppose that you are an entrepreneur who wants to develop a chain of quick-service food stores to be located in downtown parking lots. You have operated a pilot store for six months to gain operating experience and make estimates of the profitability of your idea. You now have the opportunity to acquire another location very similar to the one you already have. To make a decision, based partly on your experience with the pilot store, you feel that it is sufficient to consider four states of nature. These are S_1, the proportion of customers patronizing the parking lot who also patronize the food store is .05; S_2, the proportion is .10; S_3, the proportion is .20; and S_4, the proportion is .40. The prior probabilities that you assign to these states of nature are .25, .40, .30, and .05 respectively. The estimated discounted profits associated with opening the pilot store for each state of nature are -\$50,000, \$5,000, \$15,000, and \$100,000. The discounted profit is zero if the decision is to not open another store.

(a) Assuming the above data are as applicable to the proposed new store as to the pilot store, should you choose A_1 (open new store) or A_2 (do not open new store)?

(b) Compute EVPI.

5-5. Assume you have available the results of a survey made at the parking lot where the new store of Prob. 5-4 is to be located. The survey results are based on a probability sample and show that 15 of the 100 respondents indicated they would patronize the proposed food store.

(a) Using posterior analysis, which course of action, A_1 or A_2, should you take?

(b) If the profits associated with A_1 and A_2 can be expressed as a function of the proportion of customers using the parking lot who would also patronize the food store, what course of action would you choose if you used a classical statistical test as the basis for reaching a decision? (Let alpha risk be .05.)

(c) Compute and plot the conditional expected opportunity loss associated with the decision rule set up in part (b).

CHAPTER 6

The Normal Distribution in Bayesian Analysis

As demonstrated in Chapter 5, the analysis of a decision situation can sometimes be simplified if the probability distribution(s) involved can be expressed mathematically. The behavior of variables described by mathematical probability functions under various conditions is known, values of pertinent statistics may have been tabulated for various parameters, and so on. For this reason it may be worthwhile for the decision maker to examine the applicability of a mathematical probability function to the problem with which he is concerned.

Chapter 5 explored the use of the binomial distribution in describing the behavior of discrete variables. In some cases, however, the variable of interest may be continuous, or so numerous that we may assume that it is continuous for decision-making purposes. For example, the number of states of nature relevant to a particular situation may be so large that direct computation would be extremely time-consuming and expensive. If the decision maker were willing to act as if the probability of occurrence of states of nature in such a problem were normally distributed, his analysis could be performed much more efficiently. This chapter discusses the uses of the normal distribution in Bayesian analysis. First, the properties of the normal distribution are discussed. Next, use of the normal distribution in Bayesian prior analysis is illustrated. Then preposterior analysis is performed. Finally, sampling information for posterior analysis is integrated into the decision situation.

PROPERTIES OF THE NORMAL DISTRIBUTION

The "normal" or "Gaussian" probability distribution is often referred to as the most important distribution in statistics. The mathematical properties of this distribution were discovered by Gauss and others during their investigations of the nature of experimental errors. The pattern of distributions of these errors were so regular (for example, mean of zero, independence of observations) that they were termed "normal" errors. The normal distribution is important in Bayesian analysis for essentially three reasons: (1) it describes many real-world situations, (2) it is a limiting form of certain discrete distributions (including the binomial) which describe many real-world situations, and (3) its use simplifies calculations since published tables of its (standardized) values are readily available.

The Normal Density Function

The normal distribution is given by the probability density function

$$P_N(x) = \frac{1}{\sigma\sqrt{2\pi}} \cdot e^{-1/2\left(\frac{x-\mu}{\sigma}\right)^2} \tag{6-1}$$

where $P_N(x)$ = probability of occurrence of x, a normally distributed variable
μ = mean of the distribution
σ = standard deviation of the distribution
e = 2.7182...
π = 3.1416...

This definition generates the graph depicted in Fig. 6-1 (the reader should ignore the broken vertical lines for the moment).

The normal curve is "bell-shaped," symmetrical, and approaches the x-axis asymptotically. The distribution is specified by its mean (μ) and standard deviation (σ). If these parameters are known, $P_N(x)$ can be calculated (over some range) from Eq. 6-1. Calculations of this type prove quite burdensome, however.

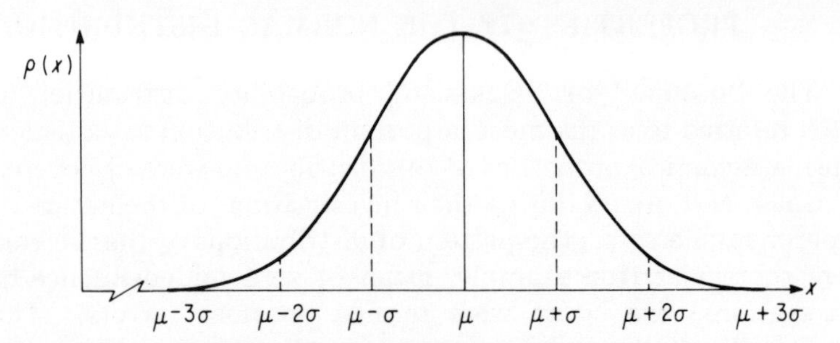

Fig. 6-1. Graph of normal density function.

The Standardized Normal Distribution

For this reason, these parameters have been "standardized." For $\mu = 0$, $\sigma = 1$, Eq. 6-1 becomes

$$P_N(x) = \frac{1}{\sqrt{2\pi}} e^{-(1/2)x^2} \tag{6-2}$$

which is the standard or unit normal density function.

Equation 6-2 contains no parameters. It describes a unique distribution, termed the *unit normal distribution*, for which values have been tabulated. Appendix B is a cumulative table of area under the right tail of the unit normal distribution. Each entry is obtained by numerically integrating Eq. 6-2 from any positive value of x to infinity.

The probability of obtaining a value greater than x from any normally distributed variable with mean μ and standard deviation σ can be obtained from Appendix B by using the transformation

$$z = \frac{x - \mu}{\sigma} \tag{6-3}$$

The probability of obtaining a value x greater than z, $P_N(x > z)$ is found by locating the entry in the body of the table that corresponds to the value for that situation. For $z = 1.00$, $P(x > z) =$ .1587; $P(x < z) = 1 - .1587 = .8413$. For $z = 1.65$, $P(x > z) =$ .04947 and $P(x < z) = 1 - .04947 = .95053$. Because the normal curve is symmetrical, the probability of x less than z standard deviations below the mean equals the probability of x greater than z standard deviations above the mean, e.g., for $z =$ 1, $P(x < z) = P(x > z) = .1587$; $P(x > z) = P(x < z) = 1 - .1587 =$ .8413.

Now the broken lines in Fig. 6-1 can be meaningfully inter-
preted. At $x = \mu + \sigma$, $z = 1.00$. From Appendix B it can be seen hat $P(\mu \leqslant x \leqslant \mu + \sigma) = .5 - P(x > z) = .5000 - .1587 = .3413$. Similarly, $P(\mu - \sigma \leqslant x \leqslant \mu) = .3413$. Consequently, 68.26 per-
cent of the area of any normal curve lies between $\pm\ \sigma$ of μ. Thus, the probability that x, a given value of a normally distributed variable, is within one standard deviation of the mean of its distribution is .6826. The corresponding probability for two standard deviations is .9545; for three standard deviations, .9973. Thus virtually any value of any normally distributed variable is within three standard deviations of the mean of the distribution.

The Central Limit Theorem

The practical importance of the normal distribution derives in large part from the central limit theorem.[1] This theorem states that

> If $x_i (i = 1, 2, \ldots, n)$ are independent random variables having the same distribution of mean μ and standard deviation σ, then as n increases without limit, the distribution of $\bar{x}$ (the mean of the x_i) approaches a normal distribution with mean μ and standard deviation $\sigma/\sqrt{n}$.

Note that the theorem does not say that $\bar{x}$ is normally distributed. But it does say that if n is sufficiently large, the distribution of $\bar{x}$ is close enough to normal to use normal tabular values in calculations involving $\bar{x}$. The magnitude of n required for "sufficiently large" is usually taken to be as small as 30.

Since the mean of the x_i from almost any type of population can be approximated by a normal distribution, the central limit theorem lends great power to the concept of sampling. If samples taken from a population can be evaluated using normal tables, the usefulness of sampling to obtain additional information is great. This point will be explored in the third section of this chapter. Before a discussion of posterior probabilities can be meaningful, however, prior probabilities must be

[1] For a more complete nonmathematical discussion, see Robert Schlaifer, *Probability and Statistics for Business Decisions* (New York: McGraw-Hill Book Company, 1959), p. 282ff; a more rigorous discussion is John E. Freund, *Mathematical Statistics* (Englewood Cliffs, N.J.: Prentice-Hall, Inc., 1962), pp. 185ff.

determined. The following section illustrates the applicability of the normal distribution to Bayesian prior analysis.

NORMAL PRIOR ANALYSIS

We shall continue the assumption introduced in Chapter 5 that only two decision alternatives (retain present facilities or expand facilities) are pertinent to our franchise problem. Consequently, the decision situation is in classic terms, a "two action problem with linear costs."[2] The payoff of each action is a function of the costs associated with each choice, and the revenues generated by the states of nature which can occur.

The Break-Even Calculation

To determine whether to retain present facilities or expand facilities, we must calculate the break-even point for expanding. The total cost of expanding is $1,000; each percentage of market share generates $500 in discounted cash flows. If he decides not to expand, our franchiser expects to obtain 6 percent of total demand. Consequently, the relevant break-even point for this situation is the point at which profit generated by retaining present facilities is the same as profit generated by expanding facilities. For the present situation, the break-even market share, denoted S_b, is

$$500S_b - 1{,}000 = 6(500)$$

$$S_b = 4{,}000/500 = 8$$

If market share exceeds 8 percent, then facilities should be expanded. If market share is less than 8 percent, retaining present facilities would be more profitable. Of course, the actual market share is the figure about which the manager is uncertain. He does not know the true state of nature.

The Sales Estimate

The manager does, however, have some knowledge, based on prior experience and judgment, about the situation. He estimates that if he expands the facility, his market share will be

[2] This convenient simplification permits us to introduce the use of the normal distribution in Bayesian analysis with a minimum of mathematical detail. More complex problems are beyond the scope of the present text.

9 percent, and he believes that there is an even chance that this estimate will be within 1 percent of the true figure.[3] If he is willing to assume that his estimates can be represented by a normal distribution, we can determine the (estimated) cumulative probability of occurrence of any given market share. Figure 6-2 presents this estimate in graphical form.

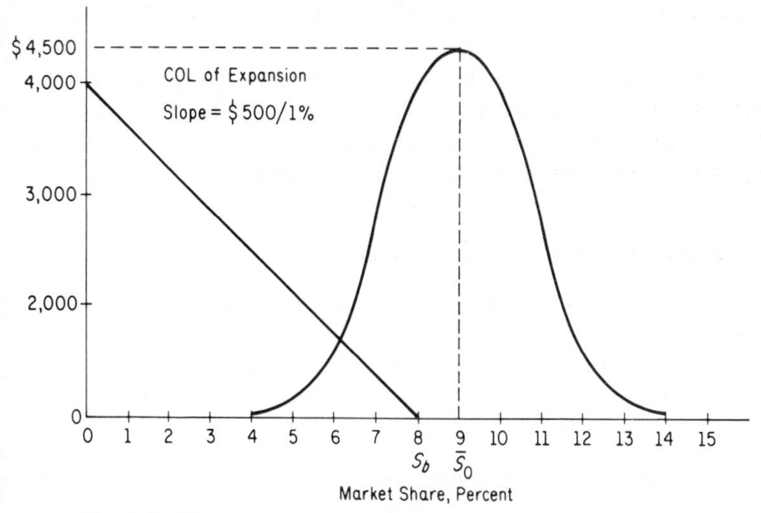

Fig. 6-2. Normally distributed market-share estimates.

Figure 6-2 was constructed in the following manner. The manager supplied four essential items of information: (1) his mean estimate of market share denoted S_0, was 9 percent, (2) he would give even odds that his estimate was within ±1 percent of the true market share, (3) he assumed that his estimates were normally distributed, and (4) COL of expansion is $500 for each market-share percentage less than 8 percent. Item (2) implies that half of the distribution falls between 8 and 10 percent of market share. As the reader can verify from Appendix B, 25 percent of the normal distribution lies between the mean and +.67 standard deviations. By symmetry, 50 percent of the distribution lies between the mean and −.67 standard

[3] Stating the manager's assumptions in this way, rather than in more elegant mathematical terms, is likely to increase understanding of the concept by operating managers.

deviations. Consequently, one percent of market share equals .67 standard deviations, so the standard deviation of the manager's estimate is

$$.67\sigma_0 = 1\%$$
$$\sigma_0 = 1.5\%$$

Since approximately 99.7 percent of the area of a normal distribution lies within $\pm 3\sigma$ of the mean, the practical implication of the above calculation is that the manager is virtually certain that market share will be at least 4.5 percent = 9.0 - 3(1.5), but no more than 13.5 percent = 9.0 + 3(1.5). The probability that market share is less than 6.0 percent = 9.0 - 2(1.5) can be read from Appendix B as .0227, and so on.

Terminal Decision from Prior Analysis

Since the probability that market share is equal to or greater than the break-even point is considerably higher than the probability of not exceeding this point, the sales manager could make the terminal decision at this stage of the analysis:

Expected present value of profit

from retaining present facilities = 6(500) = $3,000

Expected present value of profit

from expanding facilities[4] = 9(500) - 1,000 = $3,500

If the decision were to be made at this point the manager should choose to expand.

NORMAL PREPOSTERIOR ANALYSIS

Since, however, the manager's market share estimate is not perfectly reliable, there is some uncertainty associated with this

[4] For a linear function, $f = a + bx_i$, where x_i $(i = 1, \ldots, n)$ is a random variable, the mathematical expectation of the function is given by

$$E[f] = \sum_{i=1}^{n} (a + bx_i)\, P(x_i) = a \sum_{i=1}^{n} P(x_i) + b \sum_{i=1}^{n} x_i P(x_i) = a + b\bar{x}$$

decision. If market share is in fact less than 8 percent, then the better course of action is to retain the present facility. For each percentage point of market share less than 8 percent, the decision to expand the facility results in an opportunity loss of $500.

To avoid or reduce this opportunity loss, the manager might be willing to purchase additional information concerning the market share that will occur—that is, the true state of nature. Of course, if the true state is 8 percent or greater, the additional information will have no value since it will not change the decision to expand facilities. Conversely, a market share below 8 percent would indicate that retaining the present facility was the better course of action. In this case, additional information would be of value.

The value of additional information would depend on actual market share and the reliability of the information. At an opportunity loss increment of $500 per percentage point of market share, perfect information that the true market share is 7 percent (1 percent below the break-even point) would be worth $500. Perfect information that actual market share is 4 percent would be worth $500(8 - 4) = $2,000. In general, the Conditional Opportunity Loss of the decision to expand is given by

$$\text{COL} = \begin{cases} 0 \quad \text{for } S_b \leqslant S \\ \\ L(S_b - S) \quad \text{for } S_b > S \end{cases}$$

where S_b = break-even market share
S = actual market share
L = opportunity loss per percentage point of market share

Since we are dealing with uncertainty, the expected value of perfect information (EVPI) becomes the upper boundary the manager would be willing to pay for additional information. The reader will recall that EVPI is obtained by taking the expectation of the maximum conditional values for each state of nature. For the case of continuous normally distributed states of nature, the computational effort is considerably reduced by relying upon the following formula:[5]

$$\text{EVPI} = L\sigma_0\, G(D) \tag{6-4}$$

where L = opportunity loss for each percentage point of market share

σ_0 = standard deviation of the manager's estimated normal distribution of market share

$D_0 = \dfrac{|S_b - \overline{S}_0|}{\sigma_0}$, a measure of the absolute distance between the mean and the break-even point

$G(D)$ = Unit Normal Loss Integral, tabulated in Appendix C

For the present problem,

$$D_0 = \frac{|8.0 - 9.0|}{1.5} = \frac{1.0}{1.5} = .67$$

$$G(D) = .1503$$

and

$$\text{EVPI} = \$500(1.5)(.1503) = \$112.73$$

Notice that the EVPI is a function of three variables. First, the larger the opportunity loss, the larger EVPI will be. Secondly, the standard deviation of the sales estimate is a measure of the reliability of that estimate. The larger σ_0 is, the more uncertain is the estimate; consequently, additional information becomes more useful and EVPI increases as σ_0 increases. Thirdly, D_0 is a measure of the distance between the mean market share estimate and the break-even point. The greater the distance (in either direction), the more clear cut the decision becomes. Additional information becomes of more value as these two amounts become closer together. In other words, uncertainty is greatest when $S_b = S_0$. At that point, $D_0 = 0$, and $G(D)$ is maximum. Consequently, EVPI varies inversely with D.

The EVPI calculations for the present situation indicate that

[5] This formula involves the partial expectation of a normal distribution. For a more complete nonmathematical discussion, see Robert Schlaifer, *Probability and Statistics for Business Decisions* (New York: McGraw-Hill Book Company, 1959), Chaps. 18, 30; for derivations, see Howard Raiffa and Robert Schlaifer, *Applied Statistical Decision Theory* (Cambridge, Mass.: Harvard University Press, 1961), Chap. 11.

the manager would be willing to pay up to $112.73 for perfect information as to the true state of nature. Imperfect information would, of course, have less value. Given this result, the manager would probably make the terminal decision at this point. If he did request additional information, it would be obtained by sampling. Evaluation of sampling information in combination with prior information is termed *posterior analysis.*

NORMAL POSTERIOR ANALYSIS

Suppose that for the present problem a marketing research firm was commissioned to obtain additional information. The firm proceeded as follows. First, it obtained a list of all possible customers in the trade area. From this list, 50 customers were randomly selected and interviewed to ascertain the extent to which they would patronize the franchisee's place of business. This survey indicated that average market share was 7.2 percent, with a standard deviation of 3.0 percent. The manager now needs to combine this information with his prior information in order to reach a decision.

The Posterior Mean and Variance

Let us briefly review the characteristics of the information relevant to the present situation. The prior distribution is normal, with mean $\overline{S}_0 = 9.0$ and variance $\sigma_0^2 = (1.5)^2 = 2.25$. As a consequence of the central limit theorem, the mean of an unbiased sample of more than 30 observations from almost any type of population distribution will be approximately normally distributed with a mean equal to the sample mean and variance $\sigma_{\bar{x}}^2$ estimated by s^2/n, where s = sample standard deviation and n = number of observations. The relatively large n allows us to use the estimate of $\sigma_{\bar{x}}^2$ as if it were known to be the true population variance. Thus the following three conditions hold for the present problem: (1) the prior distribution of market share estimates is normal with known mean and variance, (2) the sampling distribution is normal, (3) the value of $\sigma_{\bar{x}}^2$ is known, or can be estimated.[6]

[6] A fourth condition (that the sample mean equals the population mean) is implied by the assumption that the sample is unbiased.

Under these conditions it can be shown that the posterior distribution of market share estimates is normal with mean $\bar{S}_1$ given by

$$\bar{S}_1 = \frac{\bar{S}_0(1/\sigma_0^2) + \bar{x}(1/\sigma_{\bar{x}}^2)}{1/\sigma_0^2 + 1/\sigma_{\bar{x}}^2} \tag{6-5}$$

and variance σ_1^2 given by

$$\sigma_1^2 = \frac{1}{1/\sigma_0^2 + 1/\sigma_{\bar{x}}^2} \tag{6-6}$$

where $\bar{S}_0$ = prior mean
σ_0^2 = prior variance
$\bar{x}$ = sample mean
$\sigma_{\bar{x}}^2$ = estimated population variance[7]

The reader should note that the posterior mean, $\bar{S}_1$, is a weighted average of the prior mean, $\bar{S}_0$, and the sample mean $\bar{x}$. The weights are the reciprocals of the variances of the prior and sampling distributions of market share respectively. The variance of the posterior distribution is the sum of these variances.

For the present case, the posterior mean is given by

$$\bar{S}_1 = \frac{9.0(1/2.25) + 7.2(1/.18)}{1/2.25 + 1/.18} = 7.33$$

where $\sigma_{\bar{x}}^2 = \dfrac{s^2}{n} = \dfrac{(3.0)^2}{50} = .18$

The posterior variance σ_1^2 is given by

$$\sigma_1^2 = \frac{1}{1/2.25 + 1/.18} = \frac{1}{5.999} = .167$$

The reader will note that the sample information has significantly altered the manager's prior expectations about market share. The posterior mean of 7.33 percent is below the break-even point, and is much closer to the sample mean, 7.20 percent, than to the prior mean, 9.0 percent.

[7]See Schlaifer (1959), Chaps. 30–34, and Raiffa and Schlaifer, Chap. 11.

The Quantity-of-Information Concept

As can be seen in Eq. 6-6, this result is due to the variances of the respective distributions. The variance of the sales manager's estimate, which rested on very little "hard" evidence, is much larger than the sample variance, which rests on information from 50 respondents. More formally, we define the following three quantities of information.

$I_0 = 1/\sigma_0^2$: quantity of information summarized by the prior mean

$I_{\bar{x}} = 1/\sigma_{\bar{x}}^2$: quantity of information summarized by the sample mean

$I_1 = I_0 + I_{\bar{x}}$: quantity of information summarized by the posterior mean, which takes both prior and sample information into account

These definitions explain a fact which, at first glance, may seem strange. That is, the dispersion of the posterior distribution, measured by its standard deviation σ_1, is less than the dispersion of either the prior distribution or the sampling distribution. The values are

$$\sigma_0 = 1.5\%$$

$$\sigma_{\bar{x}} = \sqrt{\sigma_{\bar{x}}^2} = .18 = .424$$

and

$$\sigma_1 = \sqrt{\sigma_1^2} = \sqrt{.167} = .409$$

This result makes sense when one reviews the quantity-of-information definitions: the posterior variance (Eq. 6-6) is the reciprocal of I_1, which is the sum of I_0 and $I_{\bar{x}}$. Consequently, this reciprocal must be smaller than either I_0 or $I_{\bar{x}}$. That is, the posterior distribution is a better estimate of the true mean market share than either the prior distribution or the sampling distribution because it is the result of more information.[8] Figure 6-3 illustrates these points.

One further methodological point is of interest. The manager's prior estimates, because of the rather large element of un-

[8] The reader is reminded that these results hold only when the three conditions listed above are met. See Schlaifer (1959), p. 444.

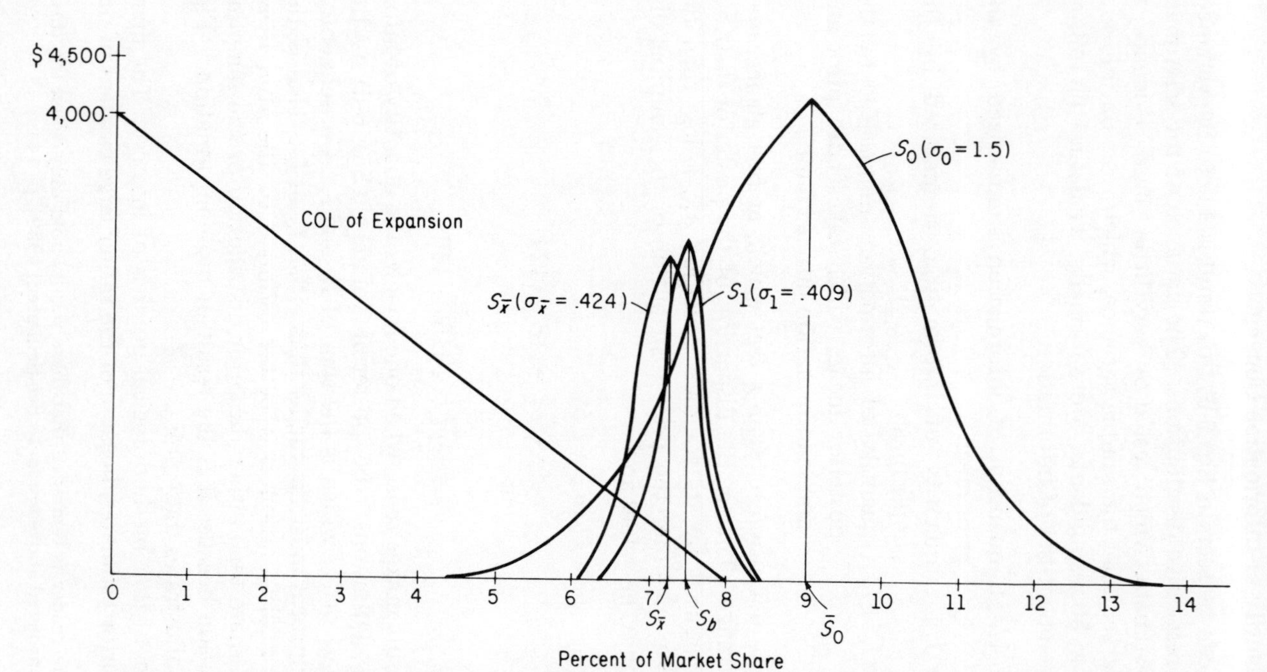

Fig. 6-3. Distribution of prior, sampling, and posterior market-share estimates.

certainty contained in them, contributed relatively little to the posterior distribution. Consequently, after receiving the sample information, the manager could have disregarded his prior estimates. Stated another way, he could set $\overline{S}_0 = 0$ and $\sigma_0^2 = \infty$, reducing Eq. 6-5 to

$$\overline{S}_1 = \overline{x} \qquad (6\text{-}7)$$

and Eq. 6-6 to

$$\sigma_1^2 = \sigma_{\overline{x}}^2 \qquad (6\text{-}8)$$

These conclusions indicate that spending a great amount of time and effort to accurately determine the decision maker's true prior distribution may not be worthwhile unless the decision maker is fairly confident of his estimates—that is, unless his estimates have a relatively small standard deviation.

The Posterior Decision

After the sample information has been evaluated, the sales manager still has three alternatives: he can retain present facilities, expand facilities, or collect additional information. The first two alternatives involve terminal action. The present value of the payoff for each action is

Retain present facilities: 6(500) = \$3,000

Expand facilities: 7.33(500) - 1,000 = \$2,665

Since the manager wishes to maximize his payoff, he should retain present facilities, provided that he makes the terminal decision at this point.

His third alternative is to gather additional information. The maximum value of this information can be calculated by modifying Eq. 6-4 to incorporate the posterior data:

$$\text{EVPI}_1 = L\sigma_1 G(D) \qquad (6\text{-}9)$$

For the present situation,

$$D_1 = \frac{|8.00 - 7.33|}{.409} = \frac{.667}{.409} = 1.63$$

and

$$\text{EVPI}_1 = 500(.409)(.02165) = \$44.27$$

Imperfect information would be worth even less. Unless he can get reliable information extremely cheap, therefore, the manager should decide at this point to retain his present facilities and consider the matter closed.

Extensions of the Analysis

The reader will recognize that the above example could be extended in several ways. First, the sample result could have been different. A sample mean of 7.8 percent, other factors unchanged, would approximately double the EVPI.

A larger sample variance would also increase EVPI. In such cases, another sample might prove worthwhile. Sequential sampling presents interesting problems, since the samples are unlikely to be independent. Secondly, a sample of size 50 was chosen arbitrarily. The problem of optimal sample size could be investigated. Thirdly, no attention was paid to sampling accuracy. Finally, the linear-cost restrictions could be relaxed. The reader interested in these topics should find the present discussion provides sufficient background for pursuing the topics in the standard references.

SUMMARY

This chapter has discussed the role of the normal distribution in Bayesian analysis. This distribution plays an important role in practical applications of statistical decision theory because it describes many real-world situations, and because its mathematical properties and published tables facilitate computations. Of course, not every situation can be described by a normal distribution, so the analyst must exercise caution in applying normal techniques. If preliminary investigation does not indicate nonnormality, however, normal techniques are useful tools for decision analysis.

REFERENCES

(See also references for Chapters 4 and 5)

Bierman, Harold, C. P. Bonini, and W. H. Houseman. *Quantitative Analysis for Business Decisions.* 3rd ed. Homewood, Ill.: Richard D. Irwin, Inc., 1968.

Extensive treatment, at an elementary level, of the role of the normal distribution in Bayesian analysis.

Raiffa, Howard, and Robert Schlaifer. *Applied Statistical Decision Theory.* Cambridge, Mass.: Harvard Graduate School of Business Administration, 1961.

This book provides proofs and underlying theory for much of Schlaifer's other books; difficult reading—advanced mathematical terminology and notation.

PROBLEMS

6-1. Reanalyze the situation in the text, given the following information:
  (a) The manager's estimated standard deviation is .25, rather than 1.5 (all other factors remain as in text).
  (b) A sample of size 100 results in sample mean of 7.8, with a sample standard deviation of 6.0 (all other factors remain as in text).

6-2. The ABC company is considering the introduction of a new product in all of its 100 outlets in a given market. Each outlet will require an expenditure of $1,750 to prepare it to handle the product. The product will sell for $4 per unit, and has a variable cost of $1.50. Demand is expected to average 800 units per outlet. This estimate is assumed to be normally distributed with a standard deviation of 150 units.
  (a) Determine the break-even volume for a single outlet.
  (b) Show that the total expected profit from introducing this product is $25,000.
  (c) Calculate the expected value of perfect information for this situation.

6-3. Suppose that for Prob. 6-1 you were given the following additional information. A sample of 50 customers interviewed in the outlets indicated on average sales of 650 units per outlet with a standard deviation of 100 units.
  (a) Show that the posterior mean and standard deviation for this situation are approximately 651 and 14 units respectively.
  (b) Calculate the posterior total expected profit.
  (c) Calculate the posterior value of perfect information.

CHAPTER 7

Summary and Managerial Perspective of the Bayesian Approach

This chapter attempts to place Bayesian analysis within the perspective of marketing management. The purpose of the chapter is to indicate how this approach fits into the marketing manager's decision-making "toolbox." First, we will briefly review the Bayesian approach to marketing decision making. The limitations of Bayesian analysis are discussed next. Then applications of Bayesian analysis are presented. Finally, some observations on the future of the Bayesian approach as a tool for marketing management are offered.

NATURE OF BAYESIAN ANALYSIS[1]

Statistical decision theory has been one of the most significant developments in management science in the past thirty years. There have been two significant theoretical contributions. First, there was the formulation of the theory of statis-

[1] This section draws from F. J. Anscombe, "Some Remarks on Bayesian Statistics," in M. W. Shelly and G. L. Bryan (eds.), *Human Judgments and Optimality* (New York: John Wiley & Sons, Inc., 1964), pp. 155–177.

tical decision (payoff) functions by Wald.[2] Second, Savage led a movement to restore respectability to the concepts of subjective probability and utility.[3] These theoretical advances resulted in the development of flexible techniques for attacking decision problems. These techniques were first introduced in textbooks by Schlaifer,[4] and by Chernoff and Moses[5] in 1959. They have been popularized and extended by many others during the past decade. Generally, the techniques used in statistical decision theory are termed *Bayesian statistics* or Bayesian analysis.

Bayesian analysis is an approach to problem solving under uncertainty. That is, the problem solver or decision maker must choose between alternative courses of action in a situation for which the actual outcome is unknown. This approach assumes that the decision maker desires to behave "rationally," that is, he wants to make consistent choices to attain some objective. This assumption implies that the decision maker has an operationally specified objective, and a criterion or criteria for measuring attainment of the objective.

The theory underlying the Bayesian approach is normative and is based on some postulates (formulated as ideals of consistency) about a decision maker's preference pattern. The theory attempts to explain the behavior of a scientifically minded scientist or administrator as he does his work. As Anscombe states, "a sort of scientific logic is desired, bearing much the same relation to the activities of scientists as mathematical logic bears to the activities of mathematicians."[6] The present theory is based on the independent work of de Finetti,[7] and von Neumann and Morgenstern.[8]

[2] A. Wald, *Statistical Decision Functions* (New York: John Wiley & Sons, Inc., 1950).

[3] L. J. Savage, *The Foundations of Statistics* (New York: John Wiley & Sons, Inc., 1954).

[4] Robert Schlaifer, *Probability and Statistics for Business Decisions* (New York: McGraw-Hill Book Company, 1959).

[5] H. Chernoff and L. E. Moses, *Elementary Decision Theory* (New York: John Wiley & Sons, Inc., 1959).

[6] Anscombe, *op. cit.*, p. 162.

[7] B. de Finetti, "La prévision: ses lois logiques, ses sources subjectives," *Annales de l'Institut Henri Poincaré*, VII (1937).

[8] J. von Neumann and O. Morgenstern, *Theory of Games and Economic Behavior*, 2d ed. (Princeton, N.J.: Princeton University Press, 1947).

The objective in most marketing problems is maximization of monetary return. The criterion, for the types of problems we have considered, is expected monetary value. As we shall illustrate in a later section, however, this criterion is not adequate for all situations involving monetary return.

The decision maker employing the Bayesian approach requires certain types of information. First, he must know or be willing to estimate the possible outcomes of the situation (states of nature), and the probability of occurrence of each. That is, he must be sufficiently knowledgeable to formulate a prior probability distribution of the various outcomes of the situation. Secondly, he must have knowledge or estimates of the payoffs (or losses) which will result from the interaction of each state and each decision alternative. Thirdly, he may be able to obtain additional information about the situation. This information takes the form of a conditional probability distribution: given the additional information, the actual outcomes or true state of nature is probabilistically specified. This distribution is usually obtained from historical records, by sampling, or by experimentation.

The third type of information is required in almost all schools of statistical decision theory. Indeed, it is the major (and in many cases only) type of information utilized in classical statistical analysis. But for some decision situations (e.g., when time is short, secrecy is essential, economic consequences are trivial, or results are clear-cut), information of this type is not obtained. To determine the value of this additional information, the second type of information, the *payoff function*, must be specified. Many classical statisticians now recognize the importance of the payoff function, and it enters, at least implicitly, into their calculations.

But the prior probability distribution is uniquely "Bayesian." This feature is the center of the major controversy between Bayesians and classical statisticians. The prior probability distribution is "subjective"—it may vary from person to person, and may even be rather imprecisely defined by a given person. However, the distribution is usually based on knowledge and experience which should not be ignored. This point is especially important in marketing decision making, since objective information is simply not available in many cases. Moreover, there are also elements of subjectivity in both payoff functions and conditional distributions of additional information. Classical statisticians appear to overlook the latter point.

In summary, we have said that Bayesian analysis is a formalized, explicit procedure for decision making under uncertainty. This procedure is particularly appropriate for marketing, because subjective judgment and economic consequences are important in marketing decision making. The value of the application of Bayesian procedures to marketing problems is threefold: (1) explicit consideration and formalization of the elements of the problem lessens the probability that relevant facts or relationships will be overlooked; (2) the procedure checks for logical consistency in the decision-making process; (3) the process can be explained by the decision maker and reviewed by others.

Bayesian analysis, however, is not a panacea for all marketing problems. The next section reviews the limitations of Bayesian analysis. Marketing managers should keep these limitations in mind when employing, or contemplating employing, Bayesian procedures.

LIMITATIONS OF BAYESIAN ANALYSIS

There are essentially four types of limitations of Bayesian analysis. These are (1) the inapplicability of expected monetary value as the criterion of choice in some decision situations, (2) difficulties inherent in the use of subjective estimates, (3) almost exponentially increasing computational difficulty as more realistic problems are analyzed, and (4) the embryonic state of development of Bayesian techniques for many practical problems. Each of these is discussed in the following paragraphs.

The EMV Criterion

Consider the following decision situation. You must choose between two alternatives: (1) you are certain to receive $25,000, and (2) the flip of a coin will determine whether you receive $100,000 (if heads appears) or nothing (tails). Which do you choose? (Please decide before reading further.)

Many people would choose the $25,000, a certain amount, to the gamble which has an expected value of

$$\text{EMV [gamble]} = [\$100{,}000, \quad 0] \begin{bmatrix} .5 \\ .5 \end{bmatrix} = \$50{,}000$$

assuming that the coin is fair.

For those people who prefer $25,000 with certainty, expected monetary value would not be an appropriate decision criterion. If you prefer the gamble, add a zero behind each figure; $250,000 for certain versus $1,000,000 or nothing, and reassess your position. At some point the consequence of the certain alternative will become too attractive to risk losing. On the other hand, if the economic consequences are sufficiently small, say $.25 for certain versus $1 or nothing, you would probably be willing to take the gamble.

The point of this little scenario is to demonstrate that expected monetary value is an appropriate decision criterion only within some range of monetary values. Where the possible consequences, in terms of economic payoff or loss, are "too good" or "too bad," no sensible decision maker will employ EMV as his criterion of choice. In these instances, he is more likely to employ *maximin*, or another criterion of choice which emphasizes risk aversion.

Ideally, the decision maker should employ *expected utility value* as his criterion of choice. In terms of utility, the person in the problem above who chooses $25,000 certain rather than a gamble with an EMV of $50,000 is said to have a "utility function" which is not linear with respect to money. The $25,000 certain offers more utility to the decision maker than does twice the amount of expected monetary return.

Bayesian analysis would be unchanged if conditional utility values were substituted for conditional payoffs. Expected utility value could then be calculated for various alternatives, and maximized. Unfortunately, as we pointed out in Chapter 3, utility is difficult to define and measure, especially when more than one person is involved.

Some work has been done in empirical utility measurement,[9] but much remains to be done. In the opinion of the

[9] F. Mosteller and P. Nogee, "An Experimental Measurement of Utility," *Journal of Political Economy*, October 1951, pp. 371–404;

authors, the development of a more comprehensive measure of value than dollars and cents would constitute a scientific breakthrough of the first order. In the meantime, marketing managers can profitably use EMV in Bayesian approaches to many problems. However, the danger inherent in undervaluing those aspects of the problem which cannot be accurately measured in monetary terms should not be ignored.

Subjective Estimates

The fact that Bayesian analysis incorporates personalistic or "subjective" information is a source of legitimate concern. Frank and Green state the problem this way:

> If the manager can only express his judgments "vaguely" or, worse yet, if he allows his general optimism (or pessimism) to influence his estimation of probabilities, subsequent computations will be suspect.[10]

Moreover, two equally knowledgeable and experienced managers may differ in the subjective estimates which they make of the same situation. There is no doubt that better decisions could be made if all necessary information were hard and objective.

While the use of subjective information constitutes a real weakness in decision making, it is not a weakness unique to Bayesian analysis. Anscombe comments, " . . . it is remarkable how ready most [classical] statisticians are to make unverified assumptions,"[11] and cites the all too rare examination of residuals in least-squares regression as an example. The point is that subjectivity cannot be eliminated from many decision problems, yet decisions must be made. The Bayesian procedure at least treats the subjective data in a logically consistent, explicit manner.

E. Edwards and L. D. Phillips, "Man as Transducer for Probabilities in Bayesian Command and Control Systems," in M. W. Shelley and G. L. Bryan (eds.) *Human Judgments and Optimality*, (New York: John Wiley & Sons, Inc., 1964); Paul E. Green, "Risk Attitudes and Chemical Investment Decisions," *Chemical Engineering Progress*, January 1963.

[10] R. E. Frank and Paul E. Green, *Quantitative Methods in Marketing* (Englewood Cliffs, N.J.: Prentice-Hall, Inc., 1967), p. 28.

[11] Anscombe, *op. cit.*, p. 163.

The Computational Burden

The more complex the problem being analyzed becomes, the greater the amount of computation required becomes. Complexity results from increasing the number of states of nature, increasing the number of alternatives, and/or decreasing the number of parameters which are assumed to be "certainty equivalents." (Payoff data, for example, could be expressed in terms of probability distributions rather than exact amounts.) Realism is often highly correlated with complexity. Consequently, the investigation of realistic problems requires methods of handling a large amount of computation.

We shall mention three ways of coping with the computational burden. First, the electronic computer can be utilized to make the required computations. Programming languages, canned routines, and time sharing terminals are increasingly available to marketing decision makers. Secondly, relationships which can be expressed mathematically can be investigated by mathematical "shortcuts." This technique was illustrated in Chapter 6 when the assumption of normality of the prior distribution of the occurrence of an infinite number of states of nature was made. Other probability distributions, (e.g., beta, Poisson, exponential, Erlang) can also be utilized in this way. Of course, use of the computer and/or higher mathematics requires some degree of expertise in these fields. Consequently, it would appear necessary to upgrade the general level of understanding of these fields among marketing managers to enable them to utilize the techniques and to communicate with experts in these fields.

The third technique for analyzing more complex problems is *simulation*. Simulation makes use of both the computer and mathematics to model the decision situation dynamically. A mathematical model of the situation is formulated, values and probability distributions are assumed for the parameters, and the model is "solved" by the computer. Thus various alternatives can be examined, and parameter values can be varied. The "sensitivity" of the optimal decision to such changes can be determined.

Limited Development of Bayesian Techniques

Perhaps its most significant limitation for marketing management is that, at the present time, Bayesian analysis is in its embryonic or at most infant stage. Although the logic of Bayesian analysis is conceptually appealing, Bayesian counterparts to many classical statistical techniques have not been developed. Quantifying the payoffs of some types of decision alternatives and determining the prior probability distributions of many situations has proven to be difficult in practice. "Canned" computer routines for many types of analyses are yet to be written, and much of the underlying theory has yet to be verified by application.

Because of these limitations, Bayesian analysis does not have the "track record," in terms of numerous applications, which classical statistical techniques have. However, researchers and decision makers are increasingly recognizing the potential power of Bayesian procedures. Consequently, applications are being reported with increasing frequency. The next section discusses some of these applications.

APPLICATIONS OF BAYESIAN ANALYSIS

Several applications of Bayesian decision procedures in marketing and other areas are presented in this section. Paul Green's work is illustrated by his now classic "Kromel" pricing problem. Rex Brown's work on measurement of sampling error, termed *error-ratio analysis*, is explained. Then several other examples are briefly reviewed. An overview of Robert Schlaifer's latest book completes this survey of Bayesian applications.

"Everclear Plastics Company"[12]

The problem in this situation concerned the pricing of a hypothetical plastics product called Kromel, used in the fabrica-

[12] From Paul E. Green, "Bayesian Decision Theory in Pricing Strategy," *Journal of Marketing*, January 1963, pp. 5–14; additional detail can be found in Robert D. Buzzell, *Mathematical Models in Marketing*

tion of upholstery for automobiles, trucks, and other commercial applications. The market for Kromel could be divided into four segments, of which the new automobile market was most significant, in that demand in this segment represented two-thirds of total potential sales volume. Kromel had not been able to penetrate this segment. An older, less durable but less expensive product, Verlon, dominated the new automobile segment of the market.

Everclear management wanted to increase Kromel profits, perhaps by penetrating the automobile segment. After some discussion, it was agreed to evaluate this objective in terms of cumulative cash flows over a five-year planning horizon. Calculations were made for cost of capital at 6 percent and at 10 percent.

Four courses of action were considered. These were (1) maintain present price of $1 per pound, (2) reduce price to $.93 per pound, (3) reduce price to $.85 per pound, and (4) reduce price to $.80 per pound. Since other marketing-mix variables, (e.g., product performance, technical assistance, and terms of sale), were virtually identical throughout the industry, it was felt that price reductions afforded the only feasible courses of action to achieve the objective of increased profits.

Profitability is a function of the interaction of the course of action selected and the market situation (state of nature) which would occur during the planning period. A number of factors had to be considered in defining relevant states of nature. First, the price cut might not result in penetration of the new automobile segment, in which case other Kromel producers might match Everclear prices in other segments, particularly if price reductions would stimulate Kromel sales in these segments. Secondly, if penetration were achieved, Verlon manufacturers might reduce the price of their product. Several possibilities (e.g., full-price match, half-price match) existed for each of these possibilities. Moreover, any of the possibilities could oc-

---

Management (Boston: Harvard Graduate School of Business Administration, 1964), pp. 112–135, and Paul E. Green and D. S. Tull, *Research for Marketing Decisions* (Englewood Cliffs, N. J.: Prentice-Hall, Inc., 1966), pp. 466–478. Although company name, and product and market data are fictitious, the structure of this problem is that of an actual marketing situation.

cur at various times during the five-year planning period, so the number of states of nature to be considered became quite large. A total of more than 400 combinations of courses of action and states of nature were evaluated.

To establish prior probability distributions for the various states of nature, several sessions were held with sales engineers, outside salesmen, and the Kromel product manager and his assistant. After some discussion, these sessions resulted in a group consensus as to the nature of each prior probability distribution.

Payoffs were calculated using the following relationship:

$$\pi [A_i] = \sum_{j=1}^{n} P(S_j) \sum_{k=1}^{5} (1 + r)^{5-k}\, T\, (D_{kj} - Z_{kj})\, (K_{kj} M_{kj})$$

where $\pi [A_i]$ = expected value of cumulative, compounded net profits, at the end of the five-year planning period, resulting from pricing strategy A_i $(i = 1, \ldots, 4)$

$P(S_j)$ = probability of occurrence of state of the jth state of nature $(j = 1, \ldots, n)$

r = annual interest rate, expressed as a decimal $(r = .06 \text{ or } .10)$

T = ratio of net to gross profits of Kromel (assumed constant in the study)

D_{kj} = Kromel price (\$/lb) in the kth year $(k = 1, \ldots, 5)$ for the jth state of nature

Z_{kj} = Kromel raw material cost (\$/lb) in the kth year for the jth state of nature $(Z_{kj} = f\,(k_{kj}, M_{kj})$, i.e., this cost was a function of Kromel sales

K_{kj} = Everclear market share of Kromel industry (expressed decimally) in the kth year for the jth state of nature

M_{kj} = total Kromel industry sales (lb) in the kth year for the jth state of nature

Computations performed on an electronic computer yielded the results summarized in Table 7-1.

The computations indicated that the \$.80/lb price would produce the greatest profit as shown in column 1. (This was the lowest price considered; it would have been interesting to com-

TABLE 7-1

Everclear Pricing Strategy Payoffs

Price Strategy, A_i	End of Period Expected Payoff (millions)	
	Variable Market Share	Constant Market Share
A_1: $1.00/lb	$26.5	$26.5
A_2: .93/lb	30.3	26.9
A_3: .85/lb	33.9	27.4
A_4: .80/lb	34.9	25.2

pare the results of a lower price.) To determine the "sensitivity" of the optimal pricing strategy, changes were made in several assumptions. The assumption that a price reduction would result in Everclear's obtaining a larger market share of the Kromel industry was found to be significant. When this assumption was changed to that of a constant market share, the best price became $.85/lb, and total profits dropped substantially, as shown in column 2 of Table 7-1. The original optimal solution was not sensitive to changes in the forecast of total industry sales (M_{kj} in the equation), nor to changes in the cost of capital (r in the equation).

This example indicates the analytical power of the Bayesian approach for decision making in complex, uncertain situations. Relevant factors were evaluated in a comprehensive, consistent manner, and the results are stated in managerial terms. The reader should also note that this was an illustration of *prior* analysis. No additional information was assembled.

"Error-Ratio Analysis"[13]

Many decisions in marketing are based upon information obtained from sampling. Consequently, the accuracy of sampling information is of significant interest to marketing decision makers. Various types of sampling errors (e.g., frame, nonresponse, response, reporting, and random sampling) are usually discussed in the literature on sampling design.[14] Unfortunately, only

[13] From Rex V. Brown, "Just How Credible Are Your Market Estimates?," *Journal of Marketing*, July 1969, pp. 46–50. A more complete discussion can be found in Rex V. Brown, *The Credibility of Estimates: A New Tool for Executives and Researchers* (Boston: Harvard Business School, Division of Research, 1969).

[14] A standard reference is M. H. Hansen, W. N. Hurwitz, and W. G. Madow, *Sample Survey Methods and Theory* (New York: John Wiley & Sons, Inc., 1953).

random-sampling error can be measured by classical statistical methods. The other types of sampling errors are subjected to physical control attempts, or are ignored. When "confidence intervals" are calculated for random-sampling error, management may mistakenly interpret these figures as probability estimates of the accuracy of the sampling information.

Brown's Bayesian approach to the determination of sampling accuracy has been termed *error-ratio analysis.* This approach involves dividing the population characteristic measured by the sample into components, determining (subjectively, or if possible, objectively) an expected value and "credible interval" (measure of dispersion) for each component, and combining these assessments to produce a probability distribution of the true value of the sampling characteristic.

To illustrate this approach, let us assume that a firm estimates its share of a particular market of 10,000 consumers by means of a mail questionnaire to 1,000 randomly selected consumers. Nine hundred usable returns show that 90 consumers favor the firm's product. This result indicates that the firm's market share is observed to be 10 percent.

But this observation is subject to possible sampling inaccuracies. We will consider three types of potential sampling error:

1. *Random sampling error* would occur if the brand share of the target population, denoted by t, differed from that of the planned sample p.
2. *Nonresponse error* would occur if the share of the planned sample p differed from that of the achieved sample a.
3. *Reporting error* would occur if r, the reported share of the achieved sample, differed from a, the true share of the achieved sample.

None of these quantities is known, so the magnitude and direction of possible errors are not known. These can, however, be expressed probabilistically.

Error-ratio analysis expresses the market-share estimate t as a function of the observed fraction r multiplied by the total-error ratio t/r:

t	=	r	$\cdot$	t/r
market share estimate		observed fraction		total-error ratio

The total-error ratio is then decomposed into the three types of errors specified above.

$\frac{t}{r} =$	$\frac{a}{r}$	$\cdot$	$\frac{p}{a}$	$\cdot$	$\frac{t}{p}$
	reporting error		nonresponse error		random-sampling error

Next, each type of error is expressed as a truth-to-estimate ratio. The decision maker, using historical, experimental, or subjective data, assigns a value to this ratio. If this value is not equal to 1.00, an error of that type is present. A measure of dispersion for this ratio is also assessed. Consequently, the market share estimate can be calculated by

$$t = r \cdot \frac{a}{r} \cdot \frac{p}{a} \cdot \frac{t}{p}$$

All of the terms on the right-hand side have been quantified.

Using this equation, a probability distribution for estimated market share t is calculated by a computer routine developed from probability theory. This distribution, expressed as an expected value and a range within which 95 percent of the distribution will fall, provides the decision maker with a logically consistent estimate, based on all available information, of his market share.

Professor Brown reports several preliminary applications of this technique, including the choice of probability vs. nonprobability sampling,[15] postevaluation of a periodic stratified sample of 20,000 businesses,[16] and a survey of parking intentions.[17]

[15] R. V. Brown and C. S. Mayer, "Towards a Rationale of Non-Probability Sampling," in P. D. Bennett, (ed.), *Marketing and Economic Development* (Chicago: American Marketing Association, 1965), pp. 295–308.

[16] C. A. B. Smith, "Personal Probability and Statistical Analysis," *Journal of the Royal Statistical Society*, Part 4 (1965), pp. 469–489.

[17] Rex V. Brown, "Evaluation of Total Survey Error," *Journal of Marketing Research*, May 1967, pp. 117–127.

)ther Bayesian Applications

Several investigators have addressed the problem of struc-uring subjective judgments and formulating prior probability listributions. Work by Winkler,[18] Smith,[19] and Brabb and Morrison[20] is of interest in this regard.

Paul Green has reported more business applications than any other investigator. In addition to the "Everclear Plastics" case, he has discussed applications in journal articles[21] and summarized two others in a book.[22] Two of Green's applications utilized prior analysis in analyzing pricing alternatives. Another used prior analysis in analyzing the size and timing of constructing additional manufacturing capacity for a new product. The other two studies dealt with analyzing the alternatives of acquiring additional information before making a terminal decision to introduce a new product. Green also participated in an interesting series of studies of information processing in simulated marketing environments.[23]

Magee has discussed an application in capital budgeting and market planning emphasizing the use of decision trees.[24] Grayson[25] and Kaufman[26] have discussed extensions of

[18] Robert L. Winkler, "The Assessment of Prior Distributions in Bayesian Analysis," *Journal of the American Statistical Association*, September 1967, pp. 776–800.

[19] Lee H. Smith, "Ranking Procedures and Subjective Probability Distributions," *Management Science*, Series B, December 1967, pp. B236–249. See also comments on this article by Paul E. Green and by D. G. Morrison, pp. B250–254.

[20] G. J. Brabb and E. D. Morrison, "The Evaluative of Subjective Information," *Journal of Marketing Research*, November 1964, pp. 40–44.

[21] "Decision Theory and Chemical Marketing," *Industrial and Engineering Chemistry*, September 1962, pp. 30–34; "Decisions Involving High Risk," *Advanced Management—Office Executive*, October 1962, pp. 18–23.

[22] Wroe Alderson and Paul E. Green, *Planning and Problem Solving in Marketing* (Homewood, Ill.: Richard D. Irwin, Inc., 1964).

[23] Paul E. Green, P. J. Robinson, and P. T. Fitzroy, *Experiments on the Value of Information in Simulated Marketing Environments* (Boston: Allyn and Bacon, Inc., 1967).

[24] John F. Magee, "Decision Trees for Decision Making," *Harvard Business Review*, July-August 1964, pp. 126–138.

[25] C. L. Grayson, Jr., *Decisions Under Uncertainty* (Boston: Harvard Business School, Division of Research, 1960).

[26] Gordon M. Kaufman, *Statistical Decision and Related Techniques in Oil and Gas Exploration* (Englewood Cliffs, N.J.: Prentice-Hall, Inc., 1963).

Bayesian analysis applied to problems in the petroleum industry. Christensen[27] has discussed the use of Bayesian analysis in evaluating bidding strategies for corporate securities. Buzzell and Slater have used Bayesian analysis to evaluate marketing strategies in the baking industry.[28] Their example was presented in a "case" format without documentation, thus it may have been either an exposition or a real-world application. Newman[29] and Rogan[30] have presented "cases" which, presumably, deal with real-world application of Bayesian analysis.

Analysis of Decisions under Uncertainty

Perhaps the best way to summarize this survey of Bayesian applications to business problems is to report on the most recent work by Robert Schlaifer. As indicated earlier, Schlaifer's book *Probability and Statistics for Business Decisions* (1959) was the first to popularize the application of Bayesian techniques to business problems. Schlaifer's continuing interest in the extension and application of Bayesian techniques to more difficult business problems is reflected in his new book *Analysis of Decisions under Uncertainty*[31] (1969). Although the two books are based on the same logical foundations, the approach and subject matter differ. The first book was concerned with the use of special techniques, formulas, charts, and tables to analyze well-defined problems. In contrast, the second book emphasizes the use of general-purpose techniques and computers to analyze large-scale, ill-defined problems involving multiple risks.

[27] C. Roland Christensen, *Strategic Aspects of Competitive Bidding for Corporate Securities* (Boston: Harvard Business School, Division of Research, 1965).

[28] Robert D. Buzzell and Charles C. Slater, "Decision Theory and Marketing Management," *Journal of Marketing*, July 1962, pp. 7–16.

[29] Joseph W. Newman, "An Application of Decision Theory under the Operating Pressures of Marketing Management," Working Paper No. 69, Graduate School of Business, Stanford University, August 1965.

[30] R. M. Rogan, "Bayesian Decision Theory as Applied to the Realignment of Salesmen's Efforts," unpublished master's thesis, Wharton School of Finance and Commerce, University of Pennsylvania, 1963. This study is summarized in Paul E. Green and D. S. Tull, *Research for Marketing Decisions* (Englewood Cliffs, N.J.: Prentice-Hall, Inc., 1966), pp. 486–493.

[31] Robert Schlaifer, *Analysis of Decisions under Uncertainty* (New York: McGraw-Hill Book Company, 1969).

Analysis of Decisions under Uncertainty is addressed to the decision maker rather than to the scholar or the technician. In the authors' opinion, this feature makes the book particularly valuable to marketing managers. Emphasis is on formulation of the decision problem, understanding the meaning and sources of objective and subjective inputs to be supplied by the decision maker, and interpretation of the results of the analysis. Special computer programs have been written and are available to reduce the computational burden of analyzing more complex problems.[32] In addition, a number of cases have been written especially for use with Schlaifer's second book.[33] The text, computer programs, and cases have been used for the past several years in a first-year course at the Harvard Business School. These materials represent perhaps the greatest effort to date to bridge the gap between theoretical developments of Bayesian techniques and applications to real-world executive decision problems.

THE FUTURE OF BAYESIAN ANALYSIS

The authors are optimistic about the future of Bayesian analysis in marketing. The approach has an underlying logic which marketing managers should find compatible: problems are structured in economic terms, the manager's judgment is utilized, and uncertainty is handled in an explicit, consistent manner. Moreover, the treatment of information as an economic good, subject to make or buy decisions, is in accord with managerial experience. Much developmental work remains, but the Bayesian approach promises to be worth the effort. In our opinion, this is an area which future marketing managers cannot afford to ignore.

Even its present state of development, Bayesian analysis has much to offer decision makers. At the very least, the logical process of the analysis sharpens managerial thinking. Moreover,

[32] *Computer Programs for a First Course in Decisions under Uncertainty* (Boston: Harvard Business School, Division of Research, forthcoming).

[33] The cases are listed in *Analysis of Decisions under Uncertainty*, p. viii, and are available from the Intercollegiate Case Clearing House, Boston, Mass.

the conceptual structuring of decision problems may lead to important insights, even if all aspects of the problem cannot be quantified. Even simplified analysis (only a few states of nature, for example) may help to "rough out" the solution to the problem. Finally, the Bayesian approach offers a means of bridging the gap between invaluable managerial experience and new, powerful quantitative techniques. Integration of these two factors will result in significant advances in marketing management. For these reasons we believe that the marketing executives and future marketing executives should be familiar with the Bayesian approach to economic decision making under uncertainty.

REFERENCES

Buzzell, Robert D. *Mathematical Models and Marketing Management.* Boston: Harvard Business School, Division of Research, 1964.

Highly readable survey of the use of mathematics in marketing, with attention to administrative problems.

Frank, R. E., and Paul E. Green. *Quantitative Methods in Marketing.* Englewood Cliffs, N.J.: Prentice-Hall, Inc., 1967.

Excellent review of the role of Bayesian analysis (and other quantitative techniques) in marketing.

Shelly, M. W., and G. L. Bryan. *Human Judgments and Optimality.* New York: John Wiley & Sons, 1964.

A collection of interesting theoretical papers on various aspects of decision theory.

APPENDIX A

Cumulative Binomial Distribution

$n = 50$

S	01	02	03	04	05	06	07	08	09	10
1	3950	6358	7819	8701	9231	9547	9734	9845	9910	9948
2	0894	2642	4447	5995	7206	8100	8735	9173	9468	9662
3	0138	0784	1892	3233	4595	3838	6892	7740	8395	8883
4	0016	0178	0628	1391	2396	3527	4673	5747	6697	7497
5	0001	0032	0168	0490	1036	1794	2710	3710	4723	5688
6		0005	0037	0144	0378	0776	1350	2081	2928	3839
7		0001	0007	0036	0118	0289	0583	1019	1596	2298
8			0001	0008	0032	0094	0220	0438	0768	1221
9				0001	0008	0027	0073	0167	0328	0579
0					0002	0007	0022	0056	0125	0245
1						0002	0006	0017	0043	0094
2							0001	0005	0013	0032
3								0001	0004	0010
4									0001	0003
5										0001

S	11	12	13	14	15	16	17	18	19	20
1	9971	9983	9991	9995	9997	9998	9999	10000	10000	10000
2	9788	9869	9920	9951	9971	9983	9990	9994	9997	9998
3	9237	9487	9661	9779	9858	9910	9944	9965	9979	9987
4	8146	8655	9042	9330	9540	9688	9792	9863	9912	9943
5	6562	7320	7956	8472	8879	9192	9428	9601	9726	9815
6	4760	5647	6463	7186	7806	8323	8741	9071	9327	9520
7	3091	3935	4789	5616	6387	7081	7686	8199	8624	8966
8	1793	2467	3217	4010	4812	5594	6328	6996	7587	8096
9	0932	1392	1955	2605	3319	4071	4832	5576	6280	6927
0	0435	0708	1074	1537	2089	2718	3403	4122	4849	5563
1	0183	0325	0535	0824	1199	1661	2203	2813	3473	4164
2	0069	0135	0242	0402	0628	0929	1309	1768	2300	2893
3	0024	0051	0100	0179	0301	0475	0714	1022	1405	1861
4	0008	0018	0037	0073	0132	0223	0357	0544	0791	1106
5	0002	0006	0013	0027	0053	0096	0164	0266	0411	0607

r \ S	11	12	13	14	15	16	17	18	19	20
16	0001	0002	0004	0009	0019	0038	0070	0120	0197	030
17			0001	0003	0007	0014	0027	0050	0087	014
18				0001	0002	0005	0010	0019	0036	006
19					0001	0001	0003	0007	0013	002
20							0001	0002	0005	000
21								0001	0002	000
22										000

r \ S	21	22	23	24	25	26	27	28	29	30
1	10000	10000	10000	10000	10000	10000	10000	10000	10000	1000
2	9999	9999	10000	10000	10000	10000	10000	10000	10000	1000
3	9992	9995	9997	9998	9999	10000	10000	10000	10000	1000
4	9964	9978	9986	9992	9995	9997	9998	9999	9999	1000
5	9877	9919	9948	9967	9979	9987	9992	9995	9997	999
6	9663	9767	9841	9893	9930	9954	9970	9981	9988	999
7	9236	9445	9603	9720	9806	9868	9911	9941	9961	997
8	8523	8874	9156	9377	9547	9676	9772	9842	9892	992
9	7505	8009	8437	8794	9084	9316	9497	9635	9740	981
10	6241	6870	7436	7934	8363	8724	9021	9260	9450	959
11	4864	5552	6210	6822	7378	7871	8299	8663	8965	921
12	3533	4201	4878	5544	6184	6782	7329	7817	8244	861
13	2383	2963	3585	4233	4890	5539	6163	6749	7287	777
14	1490	1942	2456	3023	3630	4261	4901	5534	6145	672
15	0862	1181	1565	2013	2519	3075	3669	4286	4912	553
16	0462	0665	0926	1247	1631	2075	2575	3121	3703	430
17	0229	0347	0508	0718	0983	1306	1689	2130	2623	316
18	0105	0168	0259	0384	0551	0766	1034	1359	1741	217
19	0045	0075	0122	0191	0287	0418	0590	0809	1080	140
20	0018	0031	0054	0088	0139	0212	0314	0449	0626	084
21	0006	0012	0022	0038	0063	0100	0155	0232	0338	047
22	0002	0004	0008	0015	0026	0044	0071	0112	0170	025
23	0001	0001	0003	0006	0010	0018	0031	0050	0080	012
24			0001	0002	0004	0007	0012	0021	0035	005
25				0001	0001	0002	0004	0008	0014	002
26						0001	0002	0003	0005	000
27								0001	0002	000
28									0001	000

r \ S	31	32	33	34	35	36	37	38	39	40
1	10000	10000	10000	10000	10000	10000	10000	10000	10000	100
2	10000	10000	10000	10000	10000	10000	10000	10000	10000	100
3	10000	10000	10000	10000	10000	10000	10000	10000	10000	100
4	10000	10000	10000	10000	10000	10000	10000	10000	10000	100
5	9999	9999	10000	10000	10000	10000	10000	10000	10000	100
6	9996	9997	9998	9999	9999	10000	10000	10000	10000	100
7	9984	9990	9994	9996	9998	9999	9999	10000	10000	100
8	9952	9969	9980	9987	9992	9995	9997	9998	9999	99
9	9874	9914	9942	9962	9975	9984	9990	9994	9996	99
10	9710	9794	9856	9901	9933	9955	9971	9981	9988	99

s	31	32	33	34	35	36	37	38	39	40
11	9409	9563	9683	9773	9840	9889	9924	9949	9966	9978
12	8916	9168	9371	9533	9658	9753	9825	9878	9916	9943
13	8197	8564	8873	9130	9339	9505	9635	9736	9811	9867
14	7253	7732	8157	8524	8837	9097	9310	9481	9616	9720
15	6131	6698	7223	7699	8122	8491	8805	9069	9286	9460
16	4922	5530	6120	6679	7199	7672	8094	8462	8779	9045
17	3734	4328	4931	5530	6111	6664	7179	7649	8070	8439
18	2666	3197	3760	4346	4940	5531	6105	6653	7164	7631
19	1786	2220	2703	3227	3784	4362	4949	5533	6101	6644
20	1121	1447	1826	2257	2736	3255	3805	4376	4957	5535
21	0657	0882	1156	1482	1861	2289	2764	3278	3824	4390
22	0360	0503	0685	0912	1187	1513	1890	2317	2788	3299
23	0184	0267	0379	0525	0710	0938	1214	1540	1916	2340
24	0087	0133	0196	0282	0396	0544	0730	0960	1236	1562
25	0039	0061	0094	0141	0207	0295	0411	0560	0748	0978
26	0016	0026	0042	0066	0100	0149	0216	0305	0423	0573
27	0006	0011	0018	0029	0045	0070	0106	0155	0223	0314
28	0002	0004	0007	0012	0019	0031	0048	0074	0110	0160
29	0001	0001	0002	0004	0007	0012	0020	0032	0050	0076
30			0001	0002	0003	0005	0008	0013	0021	0034
31					0001	0002	0003	0005	0008	0014
32						0001	0001	0002	0003	0005
33								0001	0001	0002
34										0001

s	41	42	43	44	45	46	47	48	49	50
1	10000	10000	10000	10000	10000	10000	10000	10000	10000	10000
2	10000	10000	10000	10000	10000	10000	10000	10000	10000	10000
3	10000	10000	10000	10000	10000	10000	10000	10000	10000	10000
4	10000	10000	10000	10000	10000	10000	10000	10000	10000	10000
5	10000	10000	10000	10000	10000	10000	10000	10000	10000	10000
6	10000	10000	10000	10000	10000	10000	10000	10000	10000	10000
7	10000	10000	10000	10000	10000	10000	10000	10000	10000	10000
8	10000	10000	10000	10000	10000	10000	10000	10000	10000	10000
9	9999	9999	10000	10000	10000	10000	10000	10000	10000	10000
10	9995	9997	9998	9999	9999	10000	10000	10000	10000	10000
11	9986	9991	9994	9997	9998	9999	9999	10000	10000	10000
12	9962	9975	9984	9990	9994	9996	9998	9999	9999	10000
13	9908	9938	9958	9973	9982	9989	9993	9996	9997	9998
14	9799	9858	9902	9933	9955	9970	9981	9988	9992	9995
15	9599	9707	9789	9851	9896	9929	9952	9968	9980	9987
16	9265	9443	9585	9696	9780	9844	9892	9926	9950	9967
17	8757	9025	9248	9429	9573	9687	9774	9839	9888	9923
18	8051	8421	8740	9010	9235	9418	9565	9680	9769	9836
19	7152	7617	8037	8406	8727	8998	9225	9410	9559	9675
20	6099	6638	7143	7608	8026	8396	8718	8991	9219	9405
21	4965	5539	6099	6635	7138	7602	8020	8391	8713	8987
22	3840	4402	4973	5543	6100	6634	7137	7599	8018	8389
23	2809	3316	3854	4412	4981	5548	6104	6636	7138	7601
24	1936	2359	2826	3331	3866	4422	4989	5554	6109	6641
25	1255	1580	1953	2375	2840	3343	3876	4431	4996	5561

r \ S	41	42	43	44	45	46	47	48	49	50
26	0762	0992	1269	1593	1966	2386	2850	3352	3885	443
27	0432	0584	0772	1003	1279	1603	1975	2395	2858	335
28	0229	0320	0439	0591	0780	1010	1286	1609	1981	239
29	0113	0164	0233	0325	0444	0595	0784	1013	1289	161
30	0052	0078	0115	0166	0235	0327	0446	0596	0784	101
31	0022	0034	0053	0079	0116	0167	0236	0327	0445	059
32	0009	0014	0022	0035	0053	0079	0116	0166	0234	032
33	0003	0005	0009	0014	0022	0035	0053	0078	0114	016
34	0001	0002	0003	0005	0009	0014	0022	0034	0052	007
35		0001	0001	0002	0003	0005	0008	0014	0021	003
36				0001	0001	0002	0003	0005	0008	001
37						0001	0001	0002	0003	000
38								0001	0001	000

$n = 100$

r \ S	01	02	03	04	05	06	07	08	09	10
1	6340	8674	9524	9831	9941	9979	9993	9998	9999	1000
2	2642	5967	8054	9128	9629	9848	9940	9977	9991	999
3	0794	3233	5802	7679	8817	9434	9742	9887	9952	998
4	0184	1410	3528	5705	7422	8570	9256	9633	9827	992
5	0034	0508	1821	3711	5640	7232	8368	9097	9526	976
6	0005	0155	0808	2116	3840	5593	7086	8201	8955	942
7	0001	0041	0312	1064	2340	3936	5557	6968	8060	882
8		0009	0106	0475	1280	2517	4012	5529	6872	793
9		0002	0032	0190	0631	1463	2660	4074	5506	679
10			0009	0068	0282	0775	1620	2780	4125	548
11			0002	0022	0115	0376	0908	1757	2882	416
12				0007	0043	0168	0469	1028	1876	297
13				0002	0015	0069	0224	0559	1138	198
14					0005	0026	0099	0282	0645	123
15					0001	0009	0041	0133	0341	072
16						0003	0016	0058	0169	039
17						0001	0006	0024	0078	020
18							0002	0009	0034	010
19							0001	0003	0014	004
20								0001	0005	002
21									0002	000
22									0001	000
23										000

r \ S	11	12	13	14	15	16	17	18	19	20
1	10000	10000	10000	10000	10000	10000	10000	10000	10000	1000
2	9999	10000	10000	10000	10000	10000	10000	10000	10000	1000
3	9992	9997	9999	10000	10000	10000	10000	10000	10000	1000
4	9966	9985	9994	9998	9999	10000	10000	10000	10000	1000
5	9886	9947	9977	9990	9996	9998	9999	10000	10000	1000
6	9698	9848	9926	9966	9984	9993	9997	9999	10000	1000
7	9328	9633	9808	9903	9953	9978	9990	9996	9998	999
8	8715	9239	9569	9766	9878	9939	9970	9986	9994	999

r \ S	11	12	13	14	15	16	17	18	19	20
9	7835	8614	9155	9508	9725	9853	9924	9962	9982	9991
10	6722	7743	8523	9078	9449	9684	9826	9908	9953	9977
11	5471	6663	7663	8440	9006	9393	9644	9800	9891	9943
12	4206	5458	6611	7591	8365	8939	9340	9605	9773	9874
13	3046	4239	5446	6566	7527	8297	8876	9289	9567	9747
14	2076	3114	4268	5436	6526	7469	8234	8819	9241	9531
15	1330	2160	3173	4294	5428	6490	7417	8177	8765	9196
16	0802	1414	2236	3227	4317	5420	6458	7370	8125	8715
17	0456	0874	1492	2305	3275	4338	5414	6429	7327	8077
18	0244	0511	0942	1563	2367	3319	4357	5408	6403	7288
19	0123	0282	0564	1006	1628	2424	3359	4374	5403	6379
20	0059	0147	0319	0614	1065	1689	2477	3395	4391	5398
21	0026	0073	0172	0356	0663	1121	1745	2525	3429	4405
22	0011	0034	0088	0196	0393	0710	1174	1797	2570	3460
23	0005	0015	0042	0103	0221	0428	0754	1223	1846	2611
24	0002	0006	0020	0051	0119	0246	0462	0796	1270	1891
25	0001	0003	0009	0024	0061	0135	0271	0496	0837	1314
26		0001	0004	0011	0030	0071	0151	0295	0528	0875
27			0001	0005	0014	0035	0081	0168	0318	0558
28			0001	0002	0006	0017	0041	0091	0184	0342
29				0001	0003	0008	0020	0048	0102	0200
30					0001	0003	0009	0024	0054	0112
31						0001	0004	0011	0027	0061
32						0001	0002	0005	0013	0031
33							0001	0002	0006	0016
34								0001	0003	0007
35									0001	0003
36										0001
37										0001

r \ S	21	22	23	24	25	26	27	28	29	30
1	10000	10000	10000	10000	10000	10000	10000	10000	10000	10000
2	10000	10000	10000	10000	10000	10000	10000	10000	10000	10000
3	10000	10000	10000	10000	10000	10000	10000	10000	10000	10000
4	10000	10000	10000	10000	10000	10000	10000	10000	10000	10000
5	10000	10000	10000	10000	10000	10000	10000	10000	10000	10000
6	10000	10000	10000	10000	10000	10000	10000	10000	10000	10000
7	10000	10000	10000	10000	10000	10000	10000	10000	10000	10000
8	9999	10000	10000	10000	10000	10000	10000	10000	10000	10000
9	9996	9998	9999	10000	10000	10000	10000	10000	10000	10000
10	9989	9995	9998	9999	10000	10000	10000	10000	10000	10000
11	9971	9986	9993	9997	9999	9999	10000	10000	10000	10000
12	9933	9965	9983	9992	9996	9998	9999	10000	10000	10000
13	9857	9922	9959	9979	9990	9995	9998	9999	10000	10000
14	9721	9840	9911	9953	9975	9988	9994	9997	9999	9999
15	9496	9695	9823	9900	9946	9972	9986	9993	9997	9998
16	9153	9462	9671	9806	9889	9939	9967	9983	9992	9996
17	8668	9112	9430	9647	9789	9878	9932	9963	9981	9990
18	8032	8625	9074	9399	9624	9773	9867	9925	9959	9978
19	7252	7991	8585	9038	9370	9601	9757	9856	9918	9955
20	6358	7220	7953	8547	9005	9342	9580	9741	9846	9911

r \ S	21	22	23	24	25	26	27	28	29	30
21	5394	6338	7189	7918	8512	8973	9316	9560	9726	983
22	4419	5391	6320	7162	7886	8479	8943	9291	9540	971
23	3488	4432	5388	6304	7136	7856	8448	8915	9267	952
24	2649	3514	4444	5386	6289	7113	7828	8420	8889	924
25	1933	2684	3539	4455	5383	6276	7091	7802	8393	88
26	1355	1972	2717	3561	4465	5381	6263	7071	7778	83
27	0911	1393	2009	2748	3583	4475	5380	6252	7053	775
28	0588	0945	1429	2043	2776	3602	4484	5378	6242	70
29	0364	0616	0978	1463	2075	2803	3621	4493	5377	62
30	0216	0386	0643	1009	1495	2105	2828	3638	4501	537
31	0123	0232	0406	0669	1038	1526	2134	2851	3654	45
32	0067	0134	0247	0427	0693	1065	1554	2160	2873	36
33	0035	0074	0144	0262	0446	0717	1091	1580	2184	28
34	0018	0039	0081	0154	0276	0465	0739	1116	1605	22
35	0009	0020	0044	0087	0164	0290	0482	0760	1139	162
36	0004	0010	0023	0048	0094	0174	0303	0499	0780	116
37	0002	0005	0011	0025	0052	0101	0183	0316	0515	07
38	0001	0002	0005	0013	0027	0056	0107	0193	0328	053
39		0001	0002	0006	0014	0030	0060	0113	0201	034
40			0001	0003	0007	0015	0032	0064	0119	021
41				0001	0003	0008	0017	0035	0068	012
42				0001	0001	0004	0008	0018	0037	007
43					0001	0002	0004	0009	0020	004
44						0001	0002	0005	0010	002
45							0001	0002	0005	001
46								0001	0002	000
47									0001	000
48										000
49										000

r \ S	31	32	33	34	35	36	37	38	39	40
1	10000	10000	10000	10000	10000	10000	10000	10000	10000	100
2	10000	10000	10000	10000	10000	10000	10000	10000	10000	100
3	10000	10000	10000	10000	10000	10000	10000	10000	10000	100
4	10000	10000	10000	10000	10000	10000	10000	10000	10000	100
5	10000	10000	10000	10000	10000	10000	10000	10000	10000	100
6	10000	10000	10000	10000	10000	10000	10000	10000	10000	100
7	10000	10000	10000	10000	10000	10000	10000	10000	10000	100
8	10000	10000	10000	10000	10000	10000	10000	10000	10000	100
9	10000	10000	10000	10000	10000	10000	10000	10000	10000	100
10	10000	10000	10000	10000	10000	10000	10000	10000	10000	100
11	10000	10000	10000	10000	10000	10000	10000	10000	10000	100
12	10000	10000	10000	10000	10000	10000	10000	10000	10000	100
13	10000	10000	10000	10000	10000	10000	10000	10000	10000	100
14	10000	10000	10000	10000	10000	10000	10000	10000	10000	100
15	9999	10000	10000	10000	10000	10000	10000	10000	10000	100
16	9998	9999	10000	10000	10000	10000	10000	10000	10000	100
17	9995	9998	9999	10000	10000	10000	10000	10000	10000	100
18	9989	9995	9997	9999	9999	10000	10000	10000	10000	100
19	9976	9988	9994	9997	9999	9999	10000	10000	10000	100
20	9950	9973	9986	9993	9997	9998	9999	10000	10000	100

s	31	32	33	34	35	36	37	38	39	40
21	9904	9946	9971	9985	9992	9996	9998	9999	10000	10000
22	9825	9898	9942	9968	9983	9991	9996	9998	9999	10000
23	9698	9816	9891	9938	9966	9982	9991	9995	9998	9999
24	9504	9685	9806	9885	9934	9963	9980	9990	9995	9997
25	9224	9487	9672	9797	9879	9930	9961	9979	9989	9994
26	8841	9204	9471	9660	9789	9873	9926	9958	9977	9988
27	8346	8820	9185	9456	9649	9780	9867	9922	9956	9976
28	7736	8325	8800	9168	9442	9638	9773	9862	9919	9954
29	7021	7717	8305	8781	9152	9429	9628	9765	9857	9916
30	6224	7007	7699	8287	8764	9137	9417	9618	9759	9852
31	5376	6216	6994	7684	8270	8748	9123	9405	9610	9752
32	4516	5376	6209	6982	7669	8254	8733	9110	9395	9602
33	3683	4523	5375	6203	6971	7656	8240	8720	9098	9385
34	2912	3696	4530	5375	6197	6961	7643	8227	8708	9087
35	2229	2929	3708	4536	5376	6192	6953	7632	8216	8697
36	1650	2249	2946	3720	4542	5376	6188	6945	7623	8205
37	1181	1671	2268	2961	3731	4547	5377	6184	6938	7614
38	0816	1200	1690	2285	2976	3741	4553	5377	6181	6932
39	0545	0833	1218	1708	2301	2989	3750	4558	5378	6178
40	0351	0558	0849	1235	1724	2316	3001	3759	4562	5379
41	0218	0361	0571	0863	1250	1739	2330	3012	3767	4567
42	0131	0226	0371	0583	0877	1265	1753	2343	3023	3775
43	0075	0136	0233	0380	0594	0889	1278	1766	2355	3033
44	0042	0079	0141	0240	0389	0605	0901	1290	1778	2365
45	0023	0044	0082	0146	0246	0397	0614	0911	1301	1789
46	0012	0024	0046	0085	0150	0252	0405	0623	0921	1311
47	0006	0012	0025	0048	0088	0154	0257	0411	0631	0930
48	0003	0006	0013	0026	0050	0091	0158	0262	0417	0638
49	0001	0003	0007	0014	0027	0052	0094	0162	0267	0423
50	0001	0001	0003	0007	0015	0029	0054	0096	0165	0271
51		0001	0002	0003	0007	0015	0030	0055	0098	0168
52			0001	0002	0004	0008	0016	0030	0056	0100
53				0001	0002	0004	0008	0016	0031	0058
54					0001	0002	0004	0008	0017	0032
55						0001	0002	0004	0009	0017
56							0001	0002	0004	0009
57								0001	0002	0004
58									0001	0002
59										0001

s	41	42	43	44	45	46	47	48	49	50
1	10000	10000	10000	10000	10000	10000	10000	10000	10000	10000
2	10000	10000	10000	10000	10000	10000	10000	10000	10000	10000
3	10000	10000	10000	10000	10000	10000	10000	10000	10000	10000
4	10000	10000	10000	10000	10000	10000	10000	10000	10000	10000
5	10000	10000	10000	10000	10000	10000	10000	10000	10000	10000
6	10000	10000	10000	10000	10000	10000	10000	10000	10000	10000
7	10000	10000	10000	10000	10000	10000	10000	10000	10000	10000
8	10000	10000	10000	10000	10000	10000	10000	10000	10000	10000
9	10000	10000	10000	10000	10000	10000	10000	10000	10000	10000
10	10000	10000	10000	10000	10000	10000	10000	10000	10000	10000

r \ S	41	42	43	44	45	46	47	48	49	50
11	10000	10000	10000	10000	10000	10000	10000	10000	10000	1000
12	10000	10000	10000	10000	10000	10000	10000	10000	10000	1000
13	10000	10000	10000	10000	10000	10000	10000	10000	10000	1000
14	10000	10000	10000	10000	10000	10000	10000	10000	10000	1000
15	10000	10000	10000	10000	10000	10000	10000	10000	10000	1000
16	10000	10000	10000	10000	10000	10000	10000	10000	10000	1000
17	10000	10000	10000	10000	10000	10000	10000	10000	10000	1000
18	10000	10000	10000	10000	10000	10000	10000	10000	10000	1000
19	10000	10000	10000	10000	10000	10000	10000	10000	10000	1000
20	10000	10000	10000	10000	10000	10000	10000	10000	10000	1000
21	10000	10000	10000	10000	10000	10000	10000	10000	10000	1000
22	10000	10000	10000	10000	10000	10000	10000	10000	10000	1000
23	10000	10000	10000	10000	10000	10000	10000	10000	10000	1000
24	9999	9999	10000	10000	10000	10000	10000	10000	10000	1000
25	9997	9999	9999	10000	10000	10000	10000	10000	10000	1000
26	9994	9997	9999	9999	10000	10000	10000	10000	10000	1000
27	9987	9994	9997	9998	9999	10000	10000	10000	10000	1000
28	9975	9987	9993	9997	9998	9999	10000	10000	10000	1000
29	9952	9974	9986	9993	9996	9998	9999	10000	10000	1000
30	9913	9950	9972	9985	9992	9996	9998	9999	10000	1000
31	9848	9910	9948	9971	9985	9992	9996	9998	9999	1000
32	9746	9844	9907	9947	9970	9984	9992	9996	9998	999
33	9594	9741	9840	9905	9945	9969	9984	9991	9996	999
34	9376	9587	9736	9837	9902	9944	9969	9983	9991	999
35	9078	9368	9581	9732	9834	9900	9942	9968	9983	999
36	8687	9069	9361	9576	9728	9831	9899	9941	9967	998
37	8196	8678	9061	9355	9571	9724	9829	9897	9941	996
38	7606	8188	8670	9054	9349	9567	9721	9827	9896	994
39	6927	7599	8181	8663	9049	9345	9563	9719	9825	989
40	6176	6922	7594	8174	8657	9044	9341	9561	9717	982
41	5380	6174	6919	7589	8169	8653	9040	9338	9558	971
42	4571	5382	6173	6916	7585	8165	8649	9037	9335	955
43	3782	4576	5383	6173	6913	7582	8162	8646	9035	933
44	3041	3788	4580	5385	6172	6912	7580	8160	8645	903
45	2375	3049	3794	4583	5387	6173	6911	7579	8159	864
46	1799	2384	3057	3799	4587	5389	6173	6911	7579	815
47	1320	1607	2391	3063	3804	4590	5391	6174	6912	757
48	0938	1328	1815	2398	3069	3809	4593	5393	6176	691
49	0644	0944	1335	1822	2404	3074	3813	4596	5395	617
50	0428	0650	0950	1341	1827	2409	3078	3816	4599	539
51	0275	0432	0655	0955	1346	1832	2413	3082	3819	460
52	0170	0278	0436	0659	0960	1350	1836	2417	3084	382
53	0102	0172	0280	0439	0662	0963	1353	1838	2419	308
54	0059	0103	0174	0282	0441	0664	0965	1355	1840	242
55	0033	0059	0104	0175	0284	0443	0666	0967	1356	184
56	0017	0033	0060	0105	0176	0285	0444	0667	0967	135
57	0009	0018	0034	0061	0106	0177	0286	0444	0667	096
58	0004	0009	0018	0034	0061	0106	0177	0286	0444	066
59	0002	0005	0009	0018	0034	0061	0106	0177	0285	044
60	0001	0002	0005	0009	0018	0034	0061	0106	0177	028
61		0001	0002	0005	0009	0018	0034	0061	0106	017
62			0001	0002	0005	0009	0018	0034	0061	010

r \ S	41	42	43	44	45	46	47	48	49	50
63				0001	0002	0005	0009	0018	0034	0060
64					0001	0002	0005	0009	0018	0033
65						0001	0002	0005	0009	0018
66							0001	0002	0004	0009
67								0001	0002	0004
68									0001	0002
69										0001

APPENDIX **B**

Cumulative Unit Normal Distribution

$P_N\ (x > z)$

z	.00	.01	.02	.03	.04	.05	.06	.07	.08	.09
0	.5000	.4960	.4920	.4880	.4840	.4801	.4761	.4721	.4681	.4641
.1	.4602	.4562	.4522	.4483	.4443	.4404	.4364	.4325	.4286	.4247
.2	.4207	.4168	.4129	.4090	.4052	.4013	.3974	.3936	.3897	.3859
.3	.3821	.3783	.3745	.3707	.3669	.3632	.3594	.3557	.3520	.3483
.4	.3446	.3409	.3372	.3336	.3300	.3264	.3228	.3192	.3156	.3121
.5	.3085	.3050	.3015	.2981	.2946	.2912	.2877	.2843	.2810	.2776
.6	.2748	.2709	.2676	.2643	.2611	.2578	.2546	.2514	.2483	.2451
.7	.2420	.2389	.2358	.2327	.2297	.2266	.2236	.2206	.2177	.2148
.8	.2119	.2090	.2061	.2033	.2005	.1977	.1949	.1922	.1894	.1867
.9	.1841	.1814	.1788	.1762	.1736	.1711	.1685	.1660	.1635	.1611
1.0	.1587	.1562	.1539	.1515	.1492	.1469	.1446	.1423	.1401	.1379
1.1	.1357	.1335	.1314	.1292	.1271	.1251	.1230	.1210	.1190	.1170
1.2	.1151	.1131	.1112	.1093	.1075	.1056	.1038	.1020	.1003	.09853
1.3	.09680	.09510	.09342	.09176	.09012	.08851	.08691	.08534	.08379	.08226
1.4	.08076	.07927	.07780	.07636	.07493	.07353	.07215	.07078	.06944	.06811

1.5	.06681	.06552	.06426	.06301	.06178	.06057	.05938	.05821	.05705	.05592
1.6	.05480	.05370	.05262	.05155	.05050	.04947	.04846	.04746	.04648	.04551
1.7	.04457	.04363	.04272	.04182	.04093	.04006	.03920	.03836	.03754	.03673
1.8	.03593	.03515	.03438	.03362	.03288	.03216	.03144	.03074	.03005	.02938
1.9	.02872	.02807	.02743	.02680	.02619	.02559	.02500	.02442	.02385	.02330
2.0	.02275	.02222	.02169	.02118	.02068	.02018	.01970	.01923	.01876	.01881
2.1	.01786	.01743	.01700	.01659	.01618	.01578	.01539	.01500	.01463	.01426
2.2	.01390	.01355	.01321	.01287	.01255	.01222	.01191	.01160	.01130	.01101
2.3	.01072	.01044	.01017	$.0^{2}9903$	$.0^{2}9642$	$.0^{2}9387$	$.0^{2}9137$	$.0^{2}8894$	$.0^{2}8656$	$.0^{2}8424$
2.4	$.0^{2}8198$	$.0^{2}7976$	$.0^{2}7760$	$.0^{2}7549$	$.0^{2}7344$	$.0^{2}7143$	$.0^{2}6947$	$.0^{2}6756$	$.0^{2}6569$	$.0^{2}6387$
2.5	$.0^{2}6210$	$.0^{2}6037$	$.0^{2}5868$	$.0^{2}5703$	$.0^{2}5543$	$.0^{2}5386$	$.0^{2}5234$	$.0^{2}5085$	$.0^{2}4940$	$.0^{2}4799$
2.6	$.0^{2}4661$	$.0^{2}4527$	$.0^{2}4396$	$.0^{2}4269$	$.0^{2}4145$	$.0^{2}4025$	$.0^{2}3907$	$.0^{2}3793$	$.0^{2}3681$	$.0^{2}3573$
2.7	$.0^{2}3467$	$.0^{2}3364$	$.0^{2}3264$	$.0^{2}3167$	$.0^{2}3072$	$.0^{2}2980$	$.0^{2}2890$	$.0^{2}2803$	$.0^{2}2718$	$.0^{2}2635$
2.8	$.0^{2}2555$	$.0^{2}2477$	$.0^{2}2401$	$.0^{2}2327$	$.0^{2}2256$	$.0^{2}2186$	$.0^{2}2118$	$.0^{2}2052$	$.0^{2}1988$	$.0^{2}1926$
2.9	$.0^{2}1866$	$.0^{2}1807$	$.0^{2}1750$	$.0^{2}1695$	$.0^{2}1641$	$.0^{2}1589$	$.0^{2}1538$	$.0^{2}1489$	$.0^{2}1441$	$.0^{2}1395$
3.0	$.0^{2}1350$	$.0^{2}1306$	$.0^{2}1264$	$.0^{2}1223$	$.0^{2}1183$	$.0^{2}1144$	$.0^{2}1107$	$.0^{2}1070$	$.0^{2}1035$	$.0^{2}1001$
3.1	$.0^{3}9676$	$.0^{3}9354$	$.0^{3}9043$	$.0^{3}8740$	$.0^{3}8447$	$.0^{3}8164$	$.0^{3}7888$	$.0^{3}7622$	$.0^{3}7364$	$.0^{3}7114$
3.2	$.0^{3}6871$	$.0^{3}6637$	$.0^{3}6410$	$.0^{3}6190$	$.0^{3}5976$	$.0^{3}5770$	$.0^{3}5571$	$.0^{3}5377$	$.0^{3}5190$	$.0^{3}5009$
3.3	$.0^{3}4834$	$.0^{3}4665$	$.0^{3}4501$	$.0^{3}4342$	$.0^{3}4189$	$.0^{3}4041$	$.0^{3}3897$	$.0^{3}3758$	$.0^{3}3624$	$.0^{3}3495$
3.4	$.0^{3}3369$	$.0^{3}3248$	$.0^{3}3131$	$.0^{3}3018$	$.0^{3}2909$	$.0^{3}2803$	$.0^{3}2701$	$.0^{3}2602$	$.0^{3}2507$	$.0^{3}2415$
3.5	$.0^{3}2326$	$.0^{3}2241$	$.0^{3}2158$	$.0^{3}2078$	$.0^{3}2001$	$.0^{3}1926$	$.0^{3}1854$	$.0^{3}1785$	$.0^{3}1718$	$.0^{3}1653$
3.6	$.0^{3}1591$	$.0^{3}1531$	$.0^{3}1473$	$.0^{3}1417$	$.0^{3}1363$	$.0^{3}1311$	$.0^{3}1261$	$.0^{3}1213$	$.0^{3}1166$	$.0^{3}1124$
3.7	$.0^{3}1078$	$.0^{3}1036$	$.0^{4}9961$	$.0^{4}9574$	$.0^{4}9201$	$.0^{4}8842$	$.0^{4}8496$	$.0^{4}8162$	$.0^{4}7844$	$.0^{4}7532$
3.8	$.0^{4}7235$	$.0^{4}6948$	$.0^{4}6673$	$.0^{4}6407$	$.0^{4}6152$	$.0^{4}5906$	$.0^{4}5669$	$.0^{4}5442$	$.0^{4}5223$	$.0^{4}5012$
3.9	$.0^{4}4810$	$.0^{4}4615$	$.0^{4}4427$	$.0^{4}4247$	$.0^{4}4074$	$.0^{4}3908$	$.0^{4}3747$	$.0^{4}3594$	$.0^{4}3446$	$.0^{4}3304$

Examples:

$$P_N(z > 3.57) = P_N(z < -3.57) = .0^{3}1785 = .0001785$$

$$P_N(z < 3.57) = P_N(z > -3.57) = 1 - .0^{3}1785 = .9998215$$

z	.00	.01	.02	.03	.04	.05	.06	.07	.08	.09
4.0	$.0^{4}3167$	$.0^{4}3036$	$.0^{4}2910$	$.0^{4}2789$	$.0^{4}2673$	$.0^{4}2561$	$.0^{4}2454$	$.0^{4}2351$	$.0^{4}2252$	$.0^{4}2457$
4.1	$.0^{4}2066$	$.0^{4}1978$	$.0^{4}1894$	$.0^{4}1814$	$.0^{4}1737$	$.0^{4}1662$	$.0^{4}1591$	$.0^{4}1523$	$.0^{4}1458$	$.0^{4}1395$
4.2	$.0^{4}1335$	$.0^{4}1277$	$.0^{4}1222$	$.0^{4}1168$	$.0^{4}1118$	$.0^{4}1069$	$.0^{4}1022$	$.0^{5}9774$	$.0^{5}9345$	$.0^{5}8934$
4.3	$.0^{5}8540$	$.0^{5}8163$	$.0^{5}7801$	$.0^{5}7455$	$.0^{5}7124$	$.0^{5}6807$	$.0^{5}6503$	$.0^{5}6212$	$.0^{5}5934$	$.0^{5}5668$
4.4	$.0^{5}5413$	$.0^{5}5169$	$.0^{5}4935$	$.0^{5}4712$	$.0^{5}4498$	$.0^{5}4294$	$.0^{5}4008$	$.0^{5}3911$	$.0^{5}3732$	$.0^{5}3561$
4.5	$.0^{5}3398$	$.0^{5}3241$	$.0^{5}3092$	$.0^{5}2949$	$.0^{5}2813$	$.0^{5}2682$	$.0^{5}2558$	$.0^{5}2439$	$.0^{5}2325$	$.0^{5}2216$
4.6	$.0^{5}2112$	$.0^{5}2013$	$.0^{5}1919$	$.0^{5}1828$	$.0^{5}1742$	$.0^{5}1660$	$.0^{5}1581$	$.0^{5}1506$	$.0^{5}1434$	$.0^{5}1366$
4.7	$.0^{5}1301$	$.0^{5}1239$	$.0^{5}1179$	$.0^{5}1123$	$.0^{5}1069$	$.0^{5}1017$	$.0^{6}9680$	$.0^{6}9211$	$.0^{6}8765$	$.0^{6}8330$
4.8	$.0^{6}7933$	$.0^{6}7547$	$.0^{6}7178$	$.0^{6}6827$	$.0^{6}6492$	$.0^{6}6173$	$.0^{6}5869$	$.0^{6}5580$	$.0^{6}5304$	$.0^{6}5042$
4.9	$.0^{6}4792$	$.0^{6}4554$	$.0^{6}4327$	$.0^{6}4111$	$.0^{6}3906$	$.0^{6}3711$	$.0^{6}3525$	$.0^{6}3348$	$.0^{6}3179$	$.0^{6}3019$

APPENDIX C

Unit Normal Loss Integral

$G(D)$

u	.00	.01	.02	.03	.04	.05	.06	.07	.08	.09
.0	.3989	.3940	.3890	.3841	.3793	.3744	.3697	.3640	.3602	.3556
.1	.3509	.3464	.3418	.3373	.3328	.3284	.3240	.3197	.3154	.3111
.2	.3069	.3027	.2986	.2944	.2904	.2863	.2824	.2784	.2745	.2706
.3	.2668	.2630	.2592	.2555	.2518	.2481	.2445	.2409	.2374	.2339
.4	.2304	.2270	.2236	.2203	.2169	.2137	.2104	.2072	.2040	.2009
.5	.1978	.1947	.1917	.1887	.1857	.1828	.1799	.1771	.1742	.1714
.6	.1687	.1659	.1633	.1606	.1580	.1554	.1528	.1503	.1478	.1453
.7	.1429	.1405	.1381	.1358	.1334	.1312	.1289	.1267	.1245	.1223
.8	.1202	.1181	.1160	.1140	.1120	.1100	.1080	.1061	.1042	.1023
.9	.1004	.09860	.09680	.09503	.09328	.09156	.08986	.08819	.08654	.08491
1.0	.08332	.08174	.08019	.07866	.07716	.07568	.07422	.07279	.07138	.06999
1.1	.06862	.06727	.06595	.06465	.06336	.06210	.06086	.05964	.05844	.05726
1.2	.05610	.05496	.05384	.05274	.05165	.05059	.04954	.04851	.04750	.04650
1.3	.04553	.04457	.04363	.04270	.04179	.04090	.04002	.03916	.03831	.03748
1.4	.03667	.03587	.03508	.03431	.03356	.03281	.03208	.03137	.03067	.02998

u	.00	.01	.02	.03	.04	.05	.06	.07	.08	.09
1.5	.02931	.02865	.02800	.02736	.02674	.02612	.02552	.02494	.02436	.02380
1.6	.02324	.02270	.02217	.02165	.02114	.02064	.02015	.01967	.01920	.01874
1.7	.01829	.01785	.01742	.01699	.01658	.01617	.01578	.01539	.01501	.01464
1.8	.01428	.01392	.01357	.01323	.01290	.01257	.01226	.01195	.01164	.01134
1.9	.01105	.01077	.01049	.01022	$.0^{2}9957$	$.0^{2}9698$	$.0^{2}9445$	$.0^{2}9198$	$.0^{2}8957$	$.0^{2}8721$
2.0	$.0^{2}8491$	$.0^{2}8266$	$.0^{2}8046$	$.0^{2}7832$	$.0^{2}7623$	$.0^{2}7418$	$.0^{2}7219$	$.0^{2}7024$	$.0^{2}6835$	$.0^{2}6649$
2.1	$.0^{2}6468$	$.0^{2}6292$	$.0^{2}6120$	$.0^{2}5952$	$.0^{2}5788$	$.0^{2}5628$	$.0^{2}5472$	$.0^{2}5320$	$.0^{2}5172$	$.0^{2}5028$
2.2	$.0^{2}4887$	$.0^{2}4750$	$.0^{2}4616$	$.0^{2}4486$	$.0^{2}4358$	$.0^{2}4235$	$.0^{2}4114$	$.0^{2}3996$	$.0^{2}3882$	$.0^{2}3770$
2.3	$.0^{2}3662$	$.0^{2}3556$	$.0^{2}3453$	$.0^{2}3352$	$.0^{2}3255$	$.0^{2}3159$	$.0^{2}3067$	$.0^{2}2977$	$.0^{2}2889$	$.0^{2}2804$
2.4	$.0^{2}2720$	$.0^{2}2640$	$.0^{2}2561$	$.0^{2}2484$	$.0^{2}2410$	$.0^{2}2337$	$.0^{2}2267$	$.0^{2}2199$	$.0^{2}2132$	$.0^{2}2067$
2.5	$.0^{2}2004$	$.0^{2}1943$	$.0^{2}1883$	$.0^{2}1826$	$.0^{2}1769$	$.0^{2}1715$	$.0^{2}1662$	$.0^{2}1610$	$.0^{2}1560$	$.0^{2}1511$
2.6	$.0^{2}1464$	$.0^{2}1418$	$.0^{2}1373$	$.0^{2}1330$	$.0^{2}1288$	$.0^{2}1247$	$.0^{2}1207$	$.0^{2}1169$	$.0^{2}1132$	$.0^{2}1095$
2.7	$.0^{2}1060$	$.0^{2}1026$	$.0^{3}9928$	$.0^{3}9607$	$.0^{3}9295$	$.0^{3}8992$	$.0^{3}8699$	$.0^{3}8414$	$.0^{3}8138$	$.0^{3}7870$
2.8	$.0^{3}7611$	$.0^{3}7359$	$.0^{3}7115$	$.0^{3}6879$	$.0^{3}6650$	$.0^{3}6428$	$.0^{3}6213$	$.0^{3}6004$	$.0^{3}5802$	$.0^{3}5606$
2.9	$.0^{3}5417$	$.0^{3}5233$	$.0^{3}5055$	$.0^{3}4883$	$.0^{3}4716$	$.0^{3}4555$	$.0^{3}4398$	$.0^{3}4247$	$.0^{3}4101$	$.0^{3}3959$
3.0	$.0^{3}3822$	$.0^{3}3689$	$.0^{3}3560$	$.0^{3}3436$	$.0^{3}3316$	$.0^{3}3199$	$.0^{3}3087$	$.0^{3}2978$	$.0^{3}2873$	$.0^{3}2771$
3.1	$.0^{3}2673$	$.0^{3}2577$	$.0^{3}2485$	$.0^{3}2396$	$.0^{3}2311$	$.0^{3}2227$	$.0^{3}2147$	$.0^{3}2070$	$.0^{3}1995$	$.0^{3}1922$
3.2	$.0^{3}1852$	$.0^{3}1785$	$.0^{3}1720$	$.0^{3}1657$	$.0^{3}1596$	$.0^{3}1537$	$.0^{3}1480$	$.0^{3}1426$	$.0^{3}1373$	$.0^{3}1322$
3.3	$.0^{3}1273$	$.0^{3}1225$	$.0^{3}1179$	$.0^{3}1135$	$.0^{3}1093$	$.0^{3}1051$	$.0^{3}1012$	$.0^{4}9734$	$.0^{4}9365$	$.0^{4}9009$
3.4	$.0^{4}8666$	$.0^{4}8335$	$.0^{4}8016$	$.0^{4}7709$	$.0^{4}7413$	$.0^{4}7127$	$.0^{4}6852$	$.0^{4}6587$	$.0^{4}6331$	$.0^{4}6085$
3.5	$.0^{4}5848$	$.0^{4}5620$	$.0^{4}5400$	$.0^{4}5188$	$.0^{4}4984$	$.0^{4}4788$	$.0^{4}4599$	$.0^{4}4417$	$.0^{4}4242$	$.0^{4}4073$
3.6	$.0^{4}3911$	$.0^{4}3755$	$.0^{4}3605$	$.0^{4}3460$	$.0^{4}3321$	$.0^{4}3188$	$.0^{4}3059$	$.0^{4}2935$	$.0^{4}2816$	$.0^{4}2702$
3.7	$.0^{4}2592$	$.0^{4}2486$	$.0^{4}2385$	$.0^{4}2287$	$.0^{4}2193$	$.0^{4}2103$	$.0^{4}2016$	$.0^{4}1933$	$.0^{4}1853$	$.0^{4}1776$
3.8	$.0^{4}1702$	$.0^{4}1632$	$.0^{4}1563$	$.0^{4}1498$	$.0^{4}1435$	$.0^{4}1375$	$.0^{4}1317$	$.0^{4}1262$	$.0^{4}1208$	$.0^{4}1157$
3.9	$.0^{4}1108$	$.0^{4}1061$	$.0^{4}1016$	$.0^{5}9723$	$.0^{5}9307$	$.0^{5}8908$	$.0^{5}8525$	$.0^{5}8158$	$.0^{5}7806$	$.0^{5}7469$
4.0	$.0^{5}7145$	$.0^{5}6835$	$.0^{5}6538$	$.0^{5}6253$	$.0^{5}5980$	$.0^{5}5718$	$.0^{5}5468$	$.0^{5}5227$	$.0^{5}4997$	$.0^{5}4777$
4.1	$.0^{5}4566$	$.0^{5}4364$	$.0^{5}4170$	$.0^{5}3985$	$.0^{5}3807$	$.0^{5}3637$	$.0^{5}3475$	$.0^{5}3319$	$.0^{5}3170$	$.0^{5}3027$
4.2	$.0^{5}2891$	$.0^{5}2760$	$.0^{5}2635$	$.0^{5}2516$	$.0^{5}2402$	$.0^{5}2292$	$.0^{5}2188$	$.0^{5}2088$	$.0^{5}1992$	$.0^{5}1901$
4.3	$.0^{5}1814$	$.0^{5}1730$	$.0^{5}1650$	$.0^{5}1574$	$.0^{5}1501$	$.0^{5}1431$	$.0^{5}1365$	$.0^{5}1301$	$.0^{5}1241$	$.0^{5}1183$

4.4	$.0^{5}1127$	$.0^{5}1074$	$.0^{5}1024$	$.0^{6}9756$	$.0^{6}9296$	$.0^{6}8857$	$.0^{6}8437$	$.0^{6}8037$	$.0^{6}7655$	$.0^{6}7290$
4.5	$.0^{6}6942$	$.0^{6}6610$	$.0^{6}6294$	$.0^{6}5992$	$.0^{6}5704$	$.0^{6}5429$	$.0^{6}5167$	$.0^{6}4917$	$.0^{6}4679$	$.0^{6}4452$
4.6	$.0^{6}4236$	$.0^{6}4029$	$.0^{6}3833$	$.0^{6}3645$	$.0^{6}3467$	$.0^{6}3297$	$.0^{6}3135$	$.0^{6}2981$	$.0^{6}2834$	$.0^{6}2694$
4.7	$.0^{6}2560$	$.0^{6}2433$	$.0^{6}2313$	$.0^{6}2197$	$.0^{6}2088$	$.0^{6}1984$	$.0^{6}1884$	$.0^{6}1790$	$.0^{6}1700$	$.0^{6}1615$
4.8	$.0^{6}1533$	$.0^{6}1456$	$.0^{6}1382$	$.0^{6}1312$	$.0^{6}1246$	$.0^{6}1182$	$.0^{6}1122$	$.0^{6}1065$	$.0^{6}1011$	$.0^{7}9588$
4.9	$.0^{7}9096$	$.0^{7}8629$	$.0^{7}8185$	$.0^{7}7763$	$.0^{7}7362$	$.0^{7}6982$	$.0^{7}6620$	$.0^{7}6276$	$.0^{7}5950$	$.0^{7}5640$

APPENDIX D

Solutions to Problems

SOLUTIONS TO PROBLEMS IN CHAPTER 2

2-1. (a) $P(A') = 1 - P(A) = 1 - .25 = .75.$

(b) $P(B') = 1 - P(B) = 1 - .30 = .70.$

(c) $P(A \cap B') = P(A) - P(A \cap B) = .25 - .15 = .10.$

(d) $P(A \cup B) = P(A) + P(B) - P(A \cap B) = .25 + .30 - .15 = .40.$

(e) $P(A' \cap B') = 1 - P(A \cup B) = 1 - .40 = .60.$

(f) $P(A' \cup B') = P(A') + P(B') - P(A' \cap B') = .75 + .70 - .60 = .85.$

2-2. (a) $P(X \neq 7) = 1 - 1/6 = 5/6.$

(b) $P(X \leqslant 4) = 1/6.$

(c) $P(X < 10) = 5/6.$

(d) $P(4 < X \leqslant 9) = 4/6.$

2-3. (a) $P(C') = 1 - P(C) = 1 - .40 = .60.$

(b) $P(M \cap C) = .12.$

(c) $P(M \cup C) = P(M) + P(C) - P(M \cap C) = .30 + .40 - .12 = .58.$

(d) $P(M \mid C) = P(M \cap C)/P(C) = .12/.40 = .30.$

(e) $P(C \mid M) = P(C \cap M)/P(M) = .12/.30 = .40.$

(f) $P(M \mid A') = P(M \cap A')/P(A') = .20/.64 = .31.$

2-4. (a)

	$P(e \mid S_i)$	$P(e' \mid S_i)$	
$P(A)$	5/30	10/30	15/30
$P(B)$	1/30	9/30	10/30
$P(C)$	1/30	4/30	5/30
	7/30	23/30	30/30

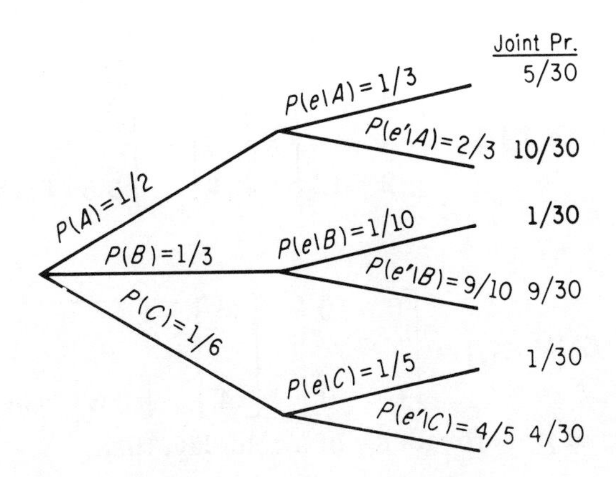

(b) $P(A|e) = P(A) \cdot P(e|A)/P(e) = (1/2 \cdot 1/3)/7/30$
$= 5/7.$

(c) Salesman B since $E[R_b] > E[R_a]$ or $E[R_c]$, i.e., $\$17 > \$15$ or $\$7$.

(d) $E[R] = (7/30) \cdot (-30) + (23/30) \cdot 60 = \$39.$

SOLUTIONS TO PROBLEMS IN CHAPTER 3

3-1. (a)

1. and 2.

	Cold	Hot	Payoffs	
			Max	Min
Coffee	\$4	2	4	2 max
Beer	-2	12	12 max	-2

3.

	Cold	Hot	Max Regret
Coffee	0	10	10
Beer	6	0	6 min

4.

$$\text{EMV }[A_i] = \begin{bmatrix} 4 & 2 \\ -2 & 12 \end{bmatrix} \cdot \begin{bmatrix} .5 \\ .5 \end{bmatrix} = \begin{bmatrix} 3 \\ 5 \end{bmatrix} \text{ beer}$$

5.

$$\text{EMV}\,[A_i] = \begin{bmatrix} 4 & 2 \\ -2 & 12 \end{bmatrix} \cdot \begin{bmatrix} .6 \\ .4 \end{bmatrix} = \begin{bmatrix} 3.2 \\ 3.6 \end{bmatrix} \text{ beer}$$

6.

$$\text{EMV}\,[A_i] = \begin{bmatrix} 0 & 10 \\ 6 & 0 \end{bmatrix} \cdot \begin{bmatrix} .6 \\ .4 \end{bmatrix} = \begin{bmatrix} 4 \\ 3.6 \end{bmatrix} \text{ beer}$$

(b) If p = probability of a cold day, then,

$$4p + 2(1 - p) = -2p + 12(1 - p)$$
$$p = 5/8$$

3-2. (a)

1. and 2.

	S_1	S_2	S_3	S_4	Payoffs Max	Min
A_1	2	3	2	5	5	2
A_2	1	4	7	1	7 max	1
A_3	2	6	5	2	6	2
A_4	3	5	3	4	5	3 max

3.

	S_1	S_2	S_3	S_4	Max Regret
A_1	1	3	5	0	5
A_2	2	2	0	4	4
A_3	1	0	2	3	3 min
A_4	0	1	4	1	4

4.

$$\text{EMV}\,[A_i] = \begin{bmatrix} 2 & 3 & 2 & 5 \\ 1 & 4 & 7 & 1 \\ 2 & 6 & 5 & 2 \\ 3 & 5 & 3 & 4 \end{bmatrix} \cdot \begin{bmatrix} .25 \\ .25 \\ .25 \\ .25 \end{bmatrix} = \begin{bmatrix} 3.0 \\ 3.25 \\ 3.75 \\ 3.75 \end{bmatrix} \begin{matrix} \\ \\ \text{max} \\ \text{max} \end{matrix}$$

5.

$$\text{EMV}[A_i] = \begin{bmatrix} 2 & 3 & 2 & 5 \\ 1 & 4 & 7 & 1 \\ 2 & 6 & 5 & 2 \\ 3 & 5 & 3 & 4 \end{bmatrix} \cdot \begin{bmatrix} .4 \\ .3 \\ .2 \\ .1 \end{bmatrix} = \begin{bmatrix} 2.6 \\ 3.1 \\ 3.8 \\ 3.7 \end{bmatrix} \begin{matrix} \\ \\ \text{max} \\ \\ \end{matrix}$$

6.

$$\text{EOL}[A_i] = \begin{bmatrix} 1 & 3 & 5 & 0 \\ 2 & 2 & 0 & 4 \\ 1 & 0 & 2 & 3 \\ 0 & 1 & 4 & 1 \end{bmatrix} \cdot \begin{bmatrix} .4 \\ .3 \\ .2 \\ .1 \end{bmatrix} = \begin{bmatrix} 2.3 \\ 1.8 \\ 1.1 \\ 1.2 \end{bmatrix} \begin{matrix} \\ \\ \text{min} \\ \\ \end{matrix}$$

3-3. (a) Conditional Payoff Matrix

Cakes Baked, A_i	Cakes Demanded, S_j						Payoffs	
	S_0	S_1	S_2	S_3	S_4	S_5	Max	Min
A_0	0	0	0	0	0	0	0	0 max
A_1	-2	1	1	1	1	1	1	-2
A_2	-4	-1	2	2	2	2	2	-4
A_3	-6	-3	0	3	3	3	3	-6
A_4	-8	-5	-2	1	4	4	4	-8
A_5	-10	-7	-4	-1	2	5	5 max	-10

Conditional Loss Matrix

Cakes Baked, A_i	Cakes Demanded, S_j						Max Regret
	S_0	S_1	S_2	S_3	S_4	S_5	
A_0	0	1	2	3	4	5	5
A_1	2	0	1	2	3	4	4 min
A_2	4	2	0	1	2	3	4 min
A_3	6	4	2	0	1	2	6
A_4	8	6	4	2	0	1	8
A_5	10	8	6	4	2	0	10

(b) From above A_5, A_0, and A_1 or A_2

(c)

$$\text{EMV}[A_i] = \begin{bmatrix} 0 & 0 & 0 & 0 & 0 & 0 \\ -2 & 1 & 1 & 1 & 1 & 1 \\ -4 & -1 & 2 & 2 & 2 & 2 \\ -6 & -3 & 0 & 3 & 3 & 3 \\ -8 & -5 & -2 & 1 & 4 & 4 \\ -10 & -7 & -4 & -1 & -1 & 5 \end{bmatrix} \cdot \begin{bmatrix} .05 \\ .10 \\ .25 \\ .30 \\ .20 \\ .10 \end{bmatrix}$$

$$= \begin{bmatrix} 0 \\ .85 \\ 1.40 \\ 1.20 \\ +.10 \\ -1.60 \end{bmatrix} \begin{matrix} \\ \\ \text{max} \\ \\ \\ \\ \end{matrix}$$

SOLUTIONS TO PROBLEMS IN CHAPTER 4

4-1. (a)

1.

$$\text{EMV}[A_i] = \begin{bmatrix} 8 & 1 \\ 4 & 3 \\ 6 & 1 \end{bmatrix} \cdot \begin{bmatrix} .3 \\ .7 \end{bmatrix} = \begin{bmatrix} 3.1 \\ 3.3 \\ 2.5 \end{bmatrix} \begin{matrix} \\ \text{max} \\ \\ \end{matrix}$$

2.

$$\text{EOL}[A_i] = \begin{bmatrix} 0 & 2 \\ 4 & 0 \\ 2 & 2 \end{bmatrix} \cdot \begin{bmatrix} .3 \\ .7 \end{bmatrix} = \begin{bmatrix} 1.4 \\ 1.2 \\ 2.0 \end{bmatrix} \begin{matrix} \\ \text{min} \\ \\ \end{matrix}$$

3.

$$\text{EPUC} = [8 \quad 3] \cdot \begin{bmatrix} .3 \\ .7 \end{bmatrix} = 4.5$$

4.

$$\text{EVPI} = 4.5 - 3.3 = 1.2 = \text{EOL}$$

(b)

$$[8 \quad 1] \cdot \begin{bmatrix} 1-p \\ p \end{bmatrix} = [4 \quad 3] \cdot \begin{bmatrix} 1-p \\ p \end{bmatrix}$$

$$p = 2/3 = P(S_2)$$

(c)

1.

	S_1	S_2	$P(R_k)$	$P(S_1 \mid R_k)$	$P(S_2 \mid R_k)$
R_1	.21	.21	.42	.500	.500
R_2	.09	.49	.58	.155	.845

2.

$$\begin{bmatrix} 8 & 1 \\ 4 & 3 \end{bmatrix} \cdot \begin{bmatrix} .500 & .155 \\ .500 & .845 \end{bmatrix} = \begin{bmatrix} 4.500 & 2.085 \\ 3.500 & 3.155 \end{bmatrix}$$

$$\text{ENGS} = [4.500 \quad 3.155] \cdot \begin{bmatrix} .42 \\ .58 \end{bmatrix} - 0.4 = 3.320$$

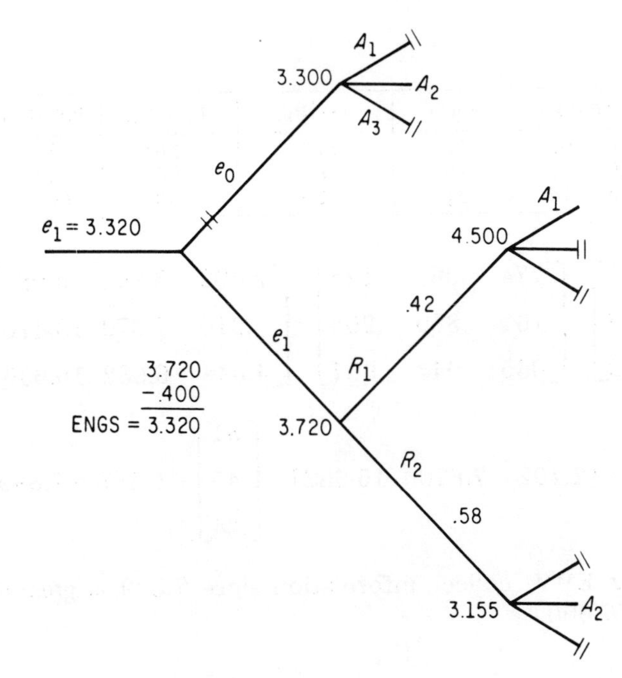

4-2. (a)

	s_{40}	s_{80}	s_{120}
A_1	2	4	6
A_2	-2	8	18
A_3	-9	7	23

(b)

$$\text{EMV}[A_i] \begin{bmatrix} 2 & 4 & 6 \\ -2 & 8 & 18 \\ -9 & 7 & 23 \end{bmatrix} \cdot \begin{bmatrix} 0.3 \\ 0.5 \\ 0.2 \end{bmatrix} = \begin{bmatrix} 3.8 \\ 7.0 \\ 5.4 \end{bmatrix} \text{max}$$

$$\text{EVPI} = \text{EVUC} - 7.0 = [2 \quad 8 \quad 23] \cdot \begin{bmatrix} 0.3 \\ 0.5 \\ 0.2 \end{bmatrix} - 7.0 = 2.2$$

Choose A_2.

(c)

	$P(R_k \mid S_j) \cdot P(S_j)$			$P(R_k)$	$P(S_1 \mid R_k)$	$P(S_2 \mid R_k)$	$P(S_3 \mid R_k)$
R_1	.24	.05	.02	.31	.774	.161	.065
R_2	.03	.40	.02	.45	.067	.889	.044
R_3	.03	.05	.16	.24	.125	.208	.667

$$\begin{bmatrix} 2 & 4 & 6 \\ -2 & 8 & 18 \\ -9 & 7 & 23 \end{bmatrix} \cdot \begin{bmatrix} .774 & .067 & .125 \\ .161 & .889 & .208 \\ .065 & .044 & .667 \end{bmatrix} = \begin{bmatrix} 2.402 & 3.994 & 5.084 \\ .910 & 7.870 & 13.470 \\ -4.344 & 6.632 & 15.636 \end{bmatrix}$$

$$\text{ENGS} = [2.402 \quad 7.870 \quad 15.636] \cdot \begin{bmatrix} .31 \\ .45 \\ .24 \end{bmatrix} - 0.150 = 7.899$$

Using EMV, collect information since 7.899 is greater than 7.000.

(d)

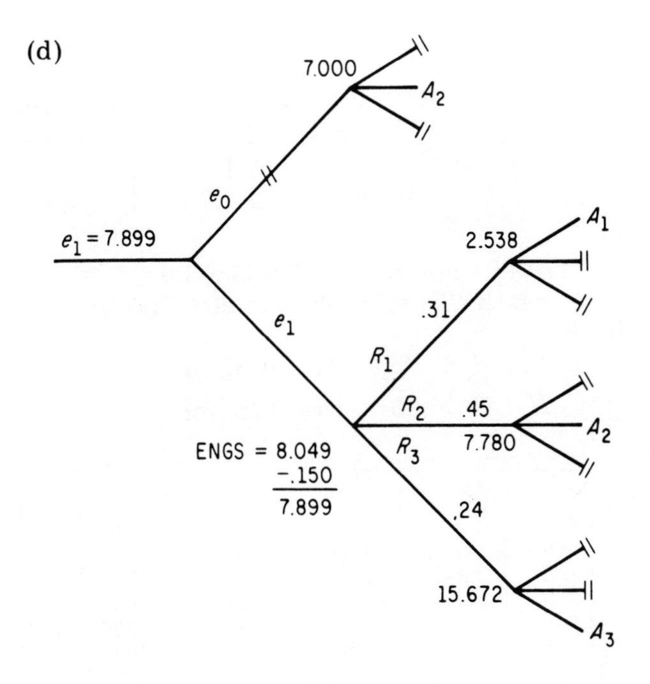

SOLUTIONS TO PROBLEMS IN CHAPTER 5

5-1. (a) Revenue = costs

$2S_b = 3000 + 2500$

$S_b = 2750$ Therefore, 2750/55000 = 5%

(b)

	s_1	s_2	s_3
A_1: Publish	−3300	0	3300
A_2: Do not publish	0	0	0

(c)

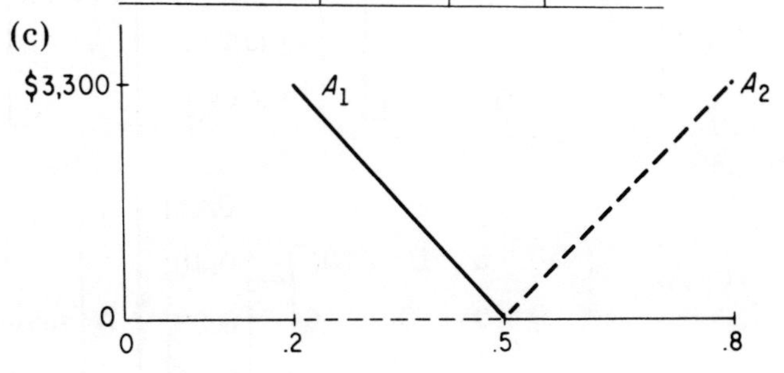

(d)

$$\text{EMV}[A_i] = \begin{bmatrix} -3300 & 0 & 3300 \\ & & \\ 0 & 0 & 0 \end{bmatrix} \cdot \begin{bmatrix} 0.1 \\ 0.3 \\ 0.6 \end{bmatrix} = \begin{bmatrix} 1650 \\ \\ 0 \end{bmatrix} \begin{matrix} \text{max} \\ \\ \\ \end{matrix}$$

5-2. (a)

$$\text{EVPI} = \min \text{EOL}[A_i] = \$330.00$$
$$\max(\text{EVSI}) = 330.00 - 74.00 = 256.00$$

(b)

$H_0: S_j \leqslant .05$ Accept H_0 for $r \leqslant 8$

$H_1: S_j > .05$ Reject H_0 for $r > 8$

S_j	Sample Result	Decision	Type Error	COL	Probability	EOL
.02	$r \leqslant 8$	A_2	none	0	.9998	0
	$r > 8$	A_1	alpha	3300	.0002	0.66
.05	$r \leqslant 8$	A_2	none	0	.9369	0
	$r > 8$	A_1	alpha	0	.0631	0
.08	$r \leqslant 8$	A_2	beta	3300	.5926	1955.58
	$r > 8$	A_1	none	0	.4074	0

5-3. (a) A_2: Do not publish

(b)

S_j	$P(S_j)$	$P(r = 4 \mid n = 100, S_j)$	$P(S_j) \cdot P(r = 4 \mid n = 100, S_j)$	$P(S_j \mid r)$
.02	.1	.0902	.00902	.09531
.05	.3	.1782	.05346	.56488
.08	.6	.0536	.03216	.33981
			.09464	1.00000

(c)

$$\text{EMV}[A_i] = \begin{bmatrix} -3300 & 0 & 3300 \\ & & \\ 0 & 0 & 0 \end{bmatrix} \cdot \begin{bmatrix} .09531 \\ .56488 \\ .33981 \end{bmatrix} = \begin{bmatrix} 812.79 \\ \\ 0 \end{bmatrix} \begin{matrix} \text{max} \\ \\ \\ \end{matrix}$$

5-4. (a)

$$\text{EMV}_0[A_i] = \begin{bmatrix} -50 & 5 & 15 & 100 \\ 0 & 0 & 0 & 0 \end{bmatrix} \cdot \begin{bmatrix} 0.25 \\ 0.40 \\ 0.30 \\ 0.05 \end{bmatrix} = \begin{bmatrix} -1 \\ \\ 0 \\ \\ \end{bmatrix} \begin{matrix} \\ \\ \text{max} \\ \\ \end{matrix}$$

(b) EVPI = EPUC − max $EMV_0[A_i]$ = 11.5 − 0 = \$11,500

5-5. (a)

S_j	$P(S_j)$	$P(r = 15 \mid n, S_j)$	$P(S_j) \cdot P(r = 15 \mid n, S_j)$	$P(S_j \mid r)$
.05	.25	.0001	.000025	.001
.10	.40	.0327	.013080	.475
.20	.30	.0481	.014430	.524
.40	.05	.0000	.000000	.000
			.027515	1.000

$$EMV_1[A_i] = \begin{bmatrix} 10{,}185 \\ 0 \end{bmatrix} \max$$

(b) If we assume the function is linear from $p = .05$ to $p = .10$, then,

$H_0: p \leqslant .095$ Accept H_0 if $r \leqslant 15$

$H_1: p > .095$ Reject H_0 if $r > 15$

and, since $r = 15$, accept H_0.

(c)

S_j	Sample Result	Decision	Type Error	COL	Probability	EOL
.05	$R \leqslant 15$	A_2	none	0	1.0000	0
	$r > 15$	A_1	alpha	50,000	0	0
.10	$r \leqslant 15$	A_2	beta	5,000	.9601	4,800
	$r > 15$	A_1	none	0	.0399	0
.20	$r \leqslant 15$	A_2	beta	15,000	.1285	1,927
	$r > 15$	A_1	none	0	.8715	0
.40	$r \leqslant 15$	A_2	beta	100,000	0	0
	$r > 15$	A_1	none	0	1.0000	0

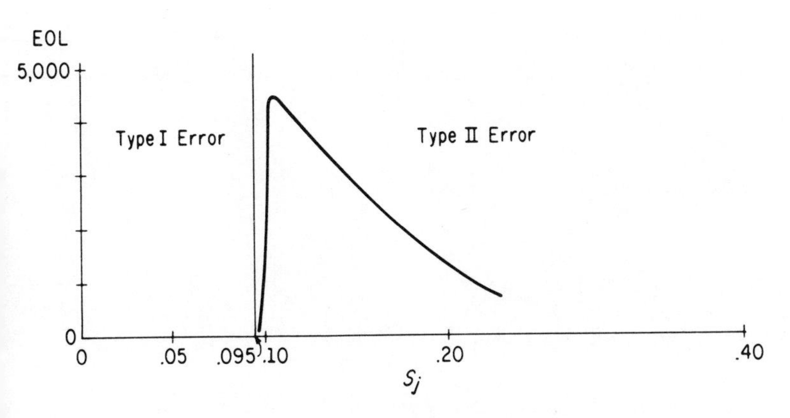

SOLUTIONS TO PROBLEMS IN CHAPTER 6

6-1. (a) $\sigma = 0.25$

$D_0 = |S_b - S|/\sigma = 1.0/.25 = 4$

$\text{EVPI}_0 = 500 \cdot .25 \cdot .000007145 = 0.$

(b) Given: $n = 100, x = 7.8, s = 6$

$$S_1 = \frac{9/0.0625 + 7.8/0.36}{1/0.0625 + 1/0.36} = 8.82$$

$\sigma_1^2 = 1/18.78 = .053$

Posterior Decision

$\text{EMV}_1 [A_1] = \$3{,}000.00$

$\text{EMV}_1 [A_2] = 8.82 \cdot 500 - 1{,}000 = \$3{,}410.00$ max

$D_1 = |8.0 - 8.82|/0.23 = 3.57$

$\text{EVPI}_1 = 500 \cdot 0.23 \cdot .00004417 = 0$

6-2. (a) $2.5S_b = 1750$

$S_b = 700$

(b) $E[P] = 100[2.5 \cdot 800 - 1{,}750] = \$25{,}000.00$

(c) $D = |800 - 700|/150 = .67$

$\text{EVPI} = 100 \cdot 2.5 \cdot 150 \cdot 0.1503 = \$5{,}636.25.$

6-3. (a) $\sigma_x^2 = s^2/n = 10{,}000/50 = 200$

$\sigma_1^2 = 22{,}500 \cdot 200/22{,}500 = 200 = 198$: $\sigma_1 = 14.07$

$$S_1 = \frac{800/22{,}500 + 650/200}{.0000444 + .005} = 651.$$

(b) $E[P] = 100[2.5 \cdot 651 - 1750] = \$12{,}200.$

(c) $D_1 = |651 - 700|/14.07 = 3.62.$

$\text{EVPI}_1 = 100 \cdot 2.5 \cdot 14.07 \cdot .00003605 = 0$

Index